HAULPAK AND LECTRA HAUL:

The World's Greatest Off-Highway Earthmoving Trucks

ERIC C. ORLEMANN

Iconografix

Iconografix
1830A Hanley Road
Hudson, Wisconsin 54016 USA

© 2012 Eric C. Orlemann

All rights reserved. No part of this work may be reproduced or used in any form by any means... graphic, electronic, or mechanical, including photocopying, recording, taping, or any other information storage and retrieval system... without written permission of the publisher.

The information in this book is true and complete to the best of our knowledge. All recommendations are made without any guarantee on the part of the author or Publisher, who also disclaim any liability incurred in connection with the use of this data or specific details.

We acknowledge that certain words, such as model names and designations, mentioned herein are the property of the trademark holder. We use them for purposes of identification only. This is not an official publication.

Iconografix books are offered at a discount when sold in quantity for promotional use. Businesses or organizations seeking details should write to the Marketing Department, Iconografix, at the above address.

Library of Congress Control Number: 2012931303

ISBN-13: 978-1-58388-292-4
ISBN-10: 1-58388-292-8

Printed in The United States of America

On the cover:
Front Top:
The mighty WABCO 3200B Haulpak pictured at the Peoria plant in August 1976.

Front Bottom:
Shown in 1979, the Unit Rig Mk36 Lectra Haul was one of the company's most popular truck lines in the 1970s.

Back:
Pictured in January 1960 is Unit Rig's prototype Lectra Haul M-64 Ore Hauler.

Book Proposals

Iconografix is a publishing company specializing in books for transportation enthusiasts. We publish in a number of different areas, including Automobiles, Auto Racing, Buses, Construction Equipment, Emergency Equipment, Farming Equipment, Railroads & Trucks. The Iconografix imprint is constantly growing and expanding into new subject areas.

Authors, editors, and knowledgeable enthusiasts in the field of transportation history are invited to contact the Editorial Department at Iconografix, 1830A Hanley Rd., Hudson, WI 54016.

Contents

Acknowledgements/Bibliography	4
Foreword	4
LeTourneau-Westinghouse and WABCO	6
Dresser and Komatsu	58
Unit Rig	78
Color Gallery	129

Acknowledgments

I have been most fortunate over the years in my business dealings to have had the pleasure of providing my photographic services to both Komatsu and Terex Mining Systems. I have worked with many fine individuals at these companies, many of which I consider good friends today.

At Komatsu, I would like to thank the following individuals for all of their help over the years in my work and book endeavors. Thanks to Richard L. Smith, Don Lindell, Sally Watkins, Kent Fales, Steve Thorson, Tom Bazzetta, and Lee Haak. I would also like to express my greatest appreciation to William E. Bontemps (formerly with Komatsu) for all of the time, effort, and help provided to me over the years.

I would also like to thank the many people at Terex (Unit Rig), both past and present, which have given me their help and time over the years. Many thanks to Larry Vargus, P.E., J. Peter Ahrenkiel, Jess Ewing, Charles V. George, and Brian Smith. I just can't say how much I will miss making my business and research trips down to the old Tulsa, Oklahoma plant, which I always looked forward to.

Thanks also to Allistair Cooke for providing additional photographic material, and Edward Creese of Rio Tinto.

Lastly, I would like to thank my good friends Keith Haddock, Dave Porter, and Urs Peyer, for always being there in my times of need. You guys are simply the best!

Eric C. Orlemann
Decatur, Illinois
December 2011

Bibliography

Gowenlock, Philip G. *WABCO Australia*. Brisbane, Queensland, Australia: Paddington Publications Pty., Ltd., 2003

Haycraft, William R. *Yellow Steel: The Story of the Earthmoving Equipment Industry*. Champaign, Illinois: University of Illinois Press, 2000.

Payne, Darwin. *Initiative in Energy: The Story of Dresser Industries*. New York, New York: Simon & Schuster, 1979.

Shelton, Jerry A. *The Unit Rig Story: An Unauthorized History of Unit Rig Equipment Co.* Self-published, Electronic, 2010.

Foreword

Post World War II, various heavy-equipment manufacturers have led the industry in the production and design of modern off-highway trucks and haulers. In the 1950s it was companies such as Euclid, Mack, and Dart that led the industry. In the 1960s, Euclid lost its way design innovation wise and would see the likes of WABCO and Unit Rig emerge as the players to beat. The 1970s was still dominated by WABCO and Unit Rig, but saw Caterpillar take great (if cautious) strides in introducing new models. The 1980s saw Caterpillar flexing its engineering muscles with the introductions of the 785 and 789 series of off-highway haulers, which would soon become leaders in their respective size classes. Companies like WISEDA and Dresser would counter Caterpillar with introductions of new designs in new payload categories, first at 220 tons then 240. Caterpillar would counter these moves with a 240 ton model of its own, the 793, in the early 1990s, firmly establishing itself as the number one producer of modern off-highway mining trucks, a status which it enjoys today, with Komatsu coming in at number two overall (but number one in the sale of 320 ton class haulers).

It was during these times that saw the birth and growth of two of the industries most respected off-highway truck lines—Haulpak and Lectra Haul. Though little known outside of the quarry and mining business worlds, they have been some of the most admired names in those industries over the past several decades. Haulpak was the trademark name of trucks built by LeTourneau-Westinghouse, WABCO, Dresser, and Komatsu. Lectra Haul was the brand name carried by haulers manufactured by Unit Rig as part of Kendavis Industries and then Terex. Both produced off-highway haulers that featured groundbreaking advancements in design and engineering for their day. So profound was the effect of these two truck lines on the industry as a whole that model lines produced by many competing manufacturers would use them as templates in the building of their hauler offerings.

The use of the "Haulpak" trade name by LeTourneau-Westinghouse was officially first introduced into use by the company on August 1, 1958. This can be confusing for some since the original LeTourneau-Westinghouse prototype off-highway truck, the LW-30, first made its public appearance in December 1956 at the company's test farm just north of Peoria. The company would continue to build a small number of haulers throughout

Shown parked at the Berkeley Pit in February 1961 is the failed Electro-Hauler concept, supposedly equipped with an experimental GE drive system.

1957, with full production commencing in early 1958. All of these early trucks should rightly be considered Haulpaks, they just didn't carry the trademark name until later in 1958. The Haulpak trade name would remain in effect throughout the various ownership name changes of the company, such as WABCO, Dresser, Komatsu Dresser, and finally Komatsu. Though primarily the Haulpak name was used to represent the company's haul trucks, it was used briefly on the giant Dresser 4000 wheel loader model starting in 1992 (originally introduced in 1991). Komatsu would quietly retire the Haulpak name in late 1999. There was no formal announcement or news release; they simply stopped placing the name on the trucks. End of story.

The "Lectra Haul" trade name was introduced by Unit Rig & Equipment Company with the introduction of their prototype M-64 Ore Hauler in January 1960. Unlike the Haulpak name which appeared 20 months after the initial unveiling of their groundbreaking prototype truck, the Lectra Haul name appeared on Unit Rig's hauler offering from day one. The Lectra Haul brand name would remain in effect until right around 1999, when it too, like Haulpak, was removed from use in advertising material and on the trucks themselves. The Unit Rig name would also be retired by Terex by the end of 2008 when it was removed from the trucks. The Unit Rig reference would still be absent during the brief time Bucyrus International was in control of the hauler line. But after Caterpillar, Inc. completed the purchase of Bucyrus in 2011, it reinstated the Unit Rig brand name to the MT-series of diesel-electric mining trucks. But sadly, the Lectra Haul trade name (like Haulpak) would remain fixtures of the Twentieth Century and not be resurrected for the Twenty-first.

It is a great achievement for a company the size of Unit Rig in the late 1950s to take on such a risky design endeavor to develop a diesel-electric powered mining hauler. R.G. LeTourneau, Inc. was the first company to build a true diesel-electric drive hauler featuring wheel motors as part of the wheel assembly themselves. Their first prototype, the TR-60 "Trolly-Dump" was completed in July 1959, just a few months before Unit Rig completed its first concept hauler to feature a GE drive system with traction wheel motors, the Lectra Haul M-64 Ore Hauler. The LeTourneau design would be up and running at Anaconda's Berkeley Pit in Butte, Montana by April 1960, right around the same time that the M-64 Lectra Haul was getting ready to be delivered to Hanna Mining for testing. It is interesting to note that at the time of the testing of the M-64, another diesel-electric drive truck utilizing an experimental General Electric drive system was taking place at Anaconda's Berkeley Pit, along with the LeTourneau TR-60. Referred to as the "Electro-Hauler," it was a 65 ton capacity, three axle unit utilizing a tractor/trailer layout. Built by a small firm in Portland, Oregon, the unit went to work in late 1960. But the Electro-Hauler was a complete disaster. It would end operations by 1962 with only one ever built. The LeTourneau TR-60 faired a little better than the Electro-Hauler concept, but not by much. Only one TR-60 would ever see the light of day. In the end Anaconda would choose a more traditional ridged-frame, rear-dump design for their truck fleet after testing a diesel-electric drive prototype M-85 Lectra Haul from Unit Rig in late 1963. Anaconda was so impressed with the M-85 that it placed orders for additional M-85 Lectra Hauls in 1964 and 1965, followed by huge orders for the higher capacity M-100 Lectra Hauls starting in 1966. Not to be left behind, WABCO would also adopt the GE motorized wheel system for its large mining trucks starting in 1965 with the introduction of the Model 120A Haulpak. The diesel-electric drive concept featuring traction wheel motors for production mining haulers was here to stay.

Chapter 1

LeTourneau-Westinghouse and WABCO

During the early 1950s, the United States government was making plans for pouring huge sums of money into new interstate expansion programs. Companies that produced earth-moving equipment were gearing up for an increase in sales and profits, while others that were not in the industry were making strategic long-term business plans that would place them into contention for some of these anticipated gains to their bottom line. One of these companies was Westinghouse Air Brake Company (WABCO). The Westinghouse Air Brake Company was originally established in 1869 by George Westinghouse, in Pittsburgh, Pennsylvania. Inventor of the railway air-brake system, which made rail travel safer and more efficient, George Westinghouse also was the founder of the Westinghouse Electric & Manufacturing Company in 1886 (renamed Westinghouse Electric Corporation in 1889). Even though both companies shared the Westinghouse name, each was an independent entity separate from each other, and not a division or subsidiary.

WABCO's first move in its expansion and diversification program of the early 1950s was its acquisition of the LeRoi Company of Sidney, Ohio, in 1952, a well-respected manufacturer of stationary and portable air compressors, air tools, and industrial drills. This company would become the LeRoi Division of Westinghouse Air Brake Company. But of far more significance was WABCO's next big purchase in 1953 of the earth-moving equipment lines of R.G. LeTourneau, Inc.

R.G. LeTourneau, Inc. was originally established in 1929 in Stockton, California, by Robert Gilmour LeTourneau, better known as R. G. LeTourneau. Considered the "dean" of modern, high speed earth-moving, LeTourneau was responsible for many industry firsts, such as equipping heavy-duty earth-moving equipment with pneumatic tires, and placing into production the first self-propelled scraper, the Tournapull. In 1935 R. G. started building an assembly plant in Peoria, Illinois, on the banks of the Illinois River, in an area known as Birkett's Hollow. Other factory sites would soon follow in Toccoa, Georgia (1939); Rydalmere, Australia

Little known in North America was the LeTourneau-Westinghouse "30" rear-dump truck, designed and built by the company's subsidiary in Australia, LeTourneau-Westinghouse Pty., Ltd. Originally produced at the Rydalmere plant in 1956, the 18-ton capacity Westinghouse 30 was purpose-built for the Australian home market (including military) and was never exported to the Americas. In 1959 the nomenclature of the little quarry truck was changed to the W18. The Westinghouse 30/W18 was never referred to as a "Haulpak" truck. Photo date is 1957. *Keith Haddock Collection*

The W18 rear-dump truck built in Australia would continue on in production until late 1967 when it was replaced with the WABCO W20. This small Australian-built, 20-ton capacity quarry hauler would only be in production until late 1970, at which time it was replaced by the model W22 of 22-ton capacity. The last W22 rolled off the Rydalmere assembly line in early 1980, bringing this line of Australian home market trucks to an end. Like the original Westinghouse 30, none of these trucks ever carried the "Haulpak" brand name. Pictured is a W20 in 1969.

The specialized "W" series construction and quarry trucks were also offered by WABCO in additional overseas marketplaces. Built at WABCO's Campinas, Brazil, manufacturing plant, the WABCO W23 was a 23-ton capacity rear-dump truck, powered by a 235-horsepower, 6-cylinder Cummins N-855-C diesel engine. This model was largely based on the W22 designed in Australia. There were plans being made by WABCO to bring this model line to the North American marketplace as the 23C Haulpak in late 1978, but financial difficulties at the company caused the project to be abandoned. Image date is 1978.

(1941); Vicksburg, Mississippi (1942); and Longview, Texas (1946). Principal products of these plants were tractor-pulled scrapers (Carryalls), self-propelled scrapers (Tournapulls), rear-dump rock trailers (Tournarockers), rubber-tired dozers (Tournadozers), dozer blades, sheep's foot rollers, and rooters (rippers). WABCO first approached LeTourneau in 1953 with an offer to purchase all of R.G. LeTourneau, Inc. But after careful analysis by WABCO accountants, it was decided that only a portion of the company would be required. In the negotiated deal between the two companies, WABCO would purchase all of LeTourneau's earth-moving equipment lines, including patents and all intellectual properties, as well as the Peoria, Toccoa, and Rydalmere assembly plants. Not included in the sale were the Longview and Vicksburg plants. The company could still continue on operating as R.G. LeTourneau, Inc., but was restricted from designing and producing commercial earth-moving related equipment for a period of five years. On May 1, 1953, the deal was finalized and a new company officially came into being as the LeTourneau-Westinghouse Company.

The new LeTourneau-Westinghouse Company (L-W) operated as a subsidiary of Westinghouse Air Brake Company, and was separate from the LeRoi Division. Though the use of the LeTourneau name was a bit confusing for customers at first, WABCO thought it was necessary to reinforce the fact that they were selling well established earth-moving equipment lines. LeTourneau-Westinghouse would continue on with expanding its earth-moving equipment offerings with the acquisition of the J.D. Adams Manufacturing Company of Indianapolis, Indiana, on January 1, 1955. This purchase included both the Indianapolis and Paris, Ontario plants. J.D. Adams was a well-known manufacturer of heavy-duty motor graders, and is credited with building the first leaning wheel grader in 1885. The Adams name would be retained on the motor grader product lines for a few years until eventually being phased out of use in the early 1960s. The Adams motor grader was a key product line that was deemed necessary by LeTourneau-Westinghouse in its quest of building and offering a credible earth-moving equipment line-up. But that still left vacancies in the track-type tractor, wheel loader, and more importantly, the rear-dump truck departments. However, that was soon about to change in regard to an off-highway rear-dump truck design.

LeTourneau-Westinghouse Company's first off-highway truck produced was a quarry-sized model designed and built at its Australian operations at Rydalmere

On a cold and snowy day in December 1956, at the company's proving grounds outside of Peoria, Illinois, LeTourneau-Westinghouse took the wraps off its new rear-dump truck model, the Model LW-30. Designed by Ralph Kress, the LW-30 was a totally new off-highway truck concept, engineered from the ground up to be the most modern quarry and mining hauler in the world. Its basic layout and engineering features were so sound that even today's off-highway mining giants still resemble it to a degree. Shown is the pilot LW-30 on December 7, 1956.

(LeTourneau-Westinghouse Pty. Ltd.) in mid-1956. Referred to as the Westinghouse 30 (renamed W18 in 1959), the spunky little 18-ton capacity rear-dump hauler was purpose-built for the Australian marketplace and never exported to the Americas. The Westinghouse 30 was a fairly simple design with a conventional frame construction, and it served well for the conditions it was designed for in Australia. But if LeTourneau-Westinghouse was going to be taken seriously as a 'real' full-line earth-moving equipment provider, it was going to need to offer a modern off-highway haul truck. And, since there were no truck manufacturers for sale at the time that met the company's needs, it would be necessary for them to build it themselves.

To make this modern truck design a reality, LeTourneau-Westinghouse enlisted the aid of a very talented engineer/consultant/manager by the name of Ralph H. Kress. Ralph Kress had served from 1950 to 1955 as the Executive Vice President/General Manager of the Dart Truck Co. of Kansas City, Missouri. In 1955 Kress left Dart and established his own company in July of that

The LeTourneau-Westinghouse Model LW-30 main engineering features would set the standard for all future quarry and mining truck designs in the industry as a whole. Key features of the truck were its triangular dump-box construction, short wheelbase, off-set cab with forward sloping windshield, and an air-hydraulic independent wheel suspension system referred to as "Hydrair." The Hydrair system eliminated the need for leaf springs commonly found in off-highway truck designs of the day, giving the LW-30 superior comfort, ride, and handling characteristics. Image taken on December 7, 1956.

The LW-30 was capable of hauling a 30-ton payload in its 20-cubic-yard dump body. Engine of choice in the pilot truck was a Cummins V-8 turbocharged diesel, rated at 375 gross horsepower. Original transmission was a sliding gear type Fuller R-1150 Roadranger, with the availability of an optional torque converter Allison CLBT 5640 Powershift. The Allison would soon be the only transmission of choice as the LW-30 approached full production status (as the LW-32). Standard tires were specified as 18.00-25, 24PR series type. Wheelbase of the pilot unit was 10 feet, 8 inches (production units would ultimately be 10 feet, 10 inches). Empty weight of the LW-30 was listed at approximately 42,320 pounds. Image date is December 19, 1956.

The LW-30 made its first public appearance at the 1957 Road Show in late January. Held at the Chicago International Amphitheater, the show was managed by the Construction Industry Manufacturers Association (CIMA), and sponsored by the American Road Builders Association. Though today many assume that the truck was identified as a "Haulpak" at this time, in fact it was not. It would not be until August 1958 that a LeTourneau-Westinghouse truck design would wear this legendary identification. But most historians give credit to the LW-30 as being the first "Haulpak," at least spiritually, if nothing else. In early 1958 the LW-30 nomenclature was changed to that of the LW-32. Image taken in March 1957.

year, Kress Automotive Engineering. The first customer of Kress was the LeTourneau-Westinghouse Company looking for input on the design and production of a state-of-the-art haul truck. LeTourneau-Westinghouse liked what Kress had to offer and took him on as a consultant whose primary duties were to design and make ready for production an off-highway truck.

Kress and LeTourneau-Westinghouse engineers worked through all of 1956 on the new North American-designed mining truck at the Peoria plant. In early December 1956 the pilot rear-dump hauler, the LW-30, was taken up to the company's 420 acre proving ground site, located approximately 12 miles north of Peoria, for initial tests and first model photography shots. The LW-30 from all accounts was a turning point in off-highway truck design, not only for LeTourneau-Westinghouse, but the industry as a whole. The LW-30 was a completely modern design from every angle, with features such as an independent, air-hydraulic suspension system referred to as "Hydrair," a "V" shaped dump body for extra capacity and a lower center of gravity, an off-set cab design, short wheelbase, and a high payload to weight ratio to name but a few. It would be the starting point for all future LeTourneau-Westinghouse mining trucks to come, as well as cement Kress' reputation as the "father" of the modern off-highway haul truck. Ralph would be offered a permanent position with LeTourneau-Westinghouse as Manager of its new truck division in 1957, which he would accept (Kress would stay with L-W through 1961, but would leave the company to go work with Caterpillar in January 1962 as Manager of Truck Development).

But development of the new LW-30 model line would not be without its ups and downs. For most of 1957, drivetrain issues kept L-W engineers busy as they fine-tuned the model design. During the development and testing of the rear-dump L-W truck, the company was also designing a tractor-trailer coal bottom-dump version based on the chassis of the original concept. In fact the first truck sold by L-W was not a rear-dump version, but the tractor-trailer variation. Referred to as the Model LW-75, the truck had the honor of being featured in Peoria's Thanksgiving Day Parade celebration which traveled down Jefferson Street. After the parade, the LW-75 was driven to Midland Electric's coal mining operation located in Farmington, which was located approximately 25 miles west of Peoria, to start a series of on-site demonstration trials for the mine. In

Originally announced in 1963, WABCO finally rolled out their first Model 25 Haulpak quarry rear dump truck in 1964. Rated with a 25-ton payload capacity, the Model 25 was powered by a 6-cylinder Cummins NT-310-CI diesel engine capable of 310 gross/285 net horsepower, or an 8-cylinder GM 8V-71N, rated at 318 gross/293 net horsepower. Standard tires were 18.00-25, 20PR series. Empty weight was listed at 42,250 pounds (42,450 with GM engine). The Model 25 was the replacement for the Model 22 which had been in the product line since 1960. Image date is May 1965.

The second rear-dump model line to be introduced by LeTourneau-Westinghouse after the initial LW-30/32 was the Model LW-27 in the summer of 1958. The original LW-27 looked much like the larger payload LW-32, including the same 10-foot, 10-inch wheelbase layout. Shown on July 9, 1958 is the pilot Model LW-27 with the "HYDRAIR" trademark on the side of the engine covers. The following month this trademark would be replaced with a new one in the form of "HAULPAK."

early 1958 Midland decided to buy the prototype coal hauler and place an order for additional units to be delivered in the second half of 1958. The first rear-dump truck models would start shipping to customers in the spring of 1958.

For most historians, the pilot LW-30 is considered the first "HAULPAK" truck model built by LeTourneau-Westinghouse. However, that is not quite correct at least as far as the use of the Haulpak trade name is concerned. The first L-W truck model to actually carry the company's new trademark name was a Model LW-27 in August 1958. This particular model was actually photographed in July of that year wearing the "HYDRAIR" trademark on the sides of its engine covers. It was re-photographed a few weeks in later in August just before final delivery, now displaying the new Haulpak identification instead of Hydrair. Documentation provided by the United States Patent and Trademark Office confirms the first use, and first use in commerce of the Haulpak trademark as August 1, 1958.

The new Haulpak trucks were well accepted in the industry and sales were very robust for LeTourneau-Westinghouse. As mine and quarry operations expanded in the 1960s, so did the size of the Haulpaks. In 1965 the company produced its first truck model featuring a diesel-electric drivetrain with General Electric traction wheel motors mounted in the rear wheel assemblies of the unit. Referred to as the Model 120A Haulpak, the truck design would lay the ground work for the next generation of diesel-electric drive haulers the company would make (under various corporate owners) for decades to come.

Starting in 1962, Westinghouse Air Brake Company started to put its 'stamp' on all its various product lines produced at its subsidiaries and various divisions in the form of an acronym of its name, WABCO. The first Haulpak trucks (along with all other L-W earth-moving equipment lines) would start being shipped with the new WABCO identification in the late summer of 1962. At this time the corporation was known as the LeTourneau-Westinghouse Company, a subsidiary of Westinghouse Air Brake Co. By early 1967 the LeTourneau name was gone from the company title, with WABCO taking over the dominant position in the corporate name game and the company now officially referred to as the Westinghouse Air Brake Company Construction Equipment Division. In late 1972 the corporate name would be modified once again to read as the WABCO Construction and Mining Equipment Group (WABCO C&MEG). By 1978 the name was shortened by one word to WABCO Construction and Mining Equipment.

After the power-steering pump was replaced with a new design, the pilot Model LW-27 was shipped into service in early August 1958. It was the company's first rear-dump truck model to wear the new "Haulpak" trademark. The LW-27 Haulpak was powered by a 6-cylinder Cummins NRTO-6-BI diesel engine, rated at 335 gross horsepower. Only transmission of choice was the automatic Allison CLBT 5640, 4-speed Torqmatic. Standard tires were 18.00-25, 20PR type. Empty hauler weight was 43,020. Payload capacity was 27 tons. Pictured on August 14, 1958, working at a rock quarry outside of Peoria, Illinois, is the pilot LW-27 wearing the new "Haulpak" truck product line identification.

The Haulpak division of mining and quarry trucks was the 'star' product line for WABCO. However, its other offerings, such as scrapers and motor graders, lagged far behind their counterparts offered by competing manufacturers. Add to this its dealer network dissatisfaction with having an incomplete line-up of equipment to offer their customers, namely no track-type tractors or wheel loaders. WABCO tried to address this situation in late 1966 by making a deal with Komatsu Ltd. of Japan to start carrying its line of track-type dozers in North America starting in January 1967. In late 1967 WABCO acquired Scoopmobile, Inc. of Portland, Oregon, primarily a builder of articulated wheel loaders. But a lack of proper management and capital concerning the new equipment offerings would actually do more harm than good. Dealers at this time were very resistant to carrying the Japanese-built dozers and investing the money into their dealerships to support them properly in the field. As a result, many Komatsu dozers were left broken down in the field with little to no service support, due to a lack of spare parts. The Komatsu and WABCO deal would quickly fall apart by the end of 1969, leaving Komatsu with a damaged reputation that would take years to erase from potential customer's minds. The Scoopmobile loader situation was a little better, but not by much. When compared to the wheel loader offerings produced by Caterpillar and International Harvester, the old Scoopmobile (the name would be dropped from use by 1970) designs were simply too antiquated to compete in such a competitive market segment. WABCO would quietly remove the loader offerings from its product line by 1974.

Westinghouse Air Brake Company had been the parent company to the LeTourneau-Westinghouse

In early 1959 LeTourneau-Westinghouse introduced a revised Model 27 Haulpak truck. It was at this time that the "LW" in the product nomenclature was officially dropped. The hauler featured a new front end and upper platform/operator step fender designs, as well as relocated engine air intake filters. All other main performance specifications remained unchanged. The Haulpak range at that time consisted of two basic rear-dump models, the 27 and 32, each utilizing the same basic 10-foot, 10-inch wheelbase and frame design. Pictured is an updated Model 27 Haulpak in November 1959.

The Model 27 Haulpak served LeTourneau-Westinghouse well in the marketplace and was offered in the product line until its replacement, the Model 30, was introduced in mid-1963. Lessons learned in the early Models 22, 27, and 32 Haulpaks would influence all of the company's truck designs for years to come. Image taken in July 1961.

Company (and WABCO Construction Equipment Division) since its creation in 1953. That would all change in 1968 over a bitter takeover battle of the company by American Standard Inc., and the Crane Company. Through various legal tactics and maneuvers between the two bidders, American Standard would come out as the victor in the buyout of the Westinghouse Air Brake Company. On June 7, 1968, WABCO officially became "An American Standard Company" subsidiary, part of the Industrial and Construction Products Division of American Standard. But it would not be smooth sailing for American Standard when it came to the construction and mining equipment lines of WABCO, as market forces and aggressive competition would severely limit the profits from these sources.

American Standard would start feeling the pressure immediately after closing the takeover deal. In the next several years it would suffer the failures of the Komatsu deal, and ultimately, the Scoopmobile wheel loader venture. In 1972, in a bid to raise more capital, it sold its LeRoi pneumatic equipment division to Dresser Industries, but it was only a stop-gap measure. Decisions made by American Standard and its subsidiary, WABCO Construction and Mining Equipment Group, would leave many of the heavy-equipment product lines in dire straits. Yet through it all, the Haulpak line of trucks continued forward with solid sales and market acceptance, especially for the larger diesel-electric drive mining haulers.

Through the 1970s, WABCO C&MG's main emphasis was on the Haulpak division and the elevating scraper product lines. Unfortunately, the standard scrapers and motor grader lines were becoming outdated and were simply out-gunned in the marketplace. WABCO waited too long to update the standard scrapers with full hydraulic controls, instead relying on the old electric motor and cable designs originally designed by R. G. LeTourneau. As for the motor graders, the industry was moving toward articulated frame offerings, something WABCO simply did not, nor intend to offer its dealers and customers. By 1980 WABCO (and American Standard) had to make some tough decisions in regard to its earth-moving equipment lines and production. The first decisions made were to end production of all scraper and motor grader models. Next was the announcement (also in 1980) of the closure of the Toccoa, Georgia plant responsible for scraper production (last unit shipped from plant inventory on March 27,

1981). Next up were the closures of the Campinas Brazil plant in mid-1982 and Rydalmere, Australia plant in December 1982. The Indianapolis, Indiana plant (which had already ended production on motor graders in 1980) was only making components and spare parts when it was finally closed in 1983. Only the Haulpak truck lines, the Peoria, Illinois plant, and Paris, Ontario, Canada facilities would survive the massive corporate cutbacks, layoffs, and shutdowns.

By this time American Standard had lost interest in its WABCO Construction and Mining Equipment division of its industrial holdings. American Standard was simply not prepared for the capital expenditures required to see the division through the tough economic times of the early 1980s due to the worldwide economic recession. In late 1983 American Standard started looking for a possible buyer for its WABCO Construction and Mining Equipment division, which now only consisted of the Haulpak lines of mining trucks. In 1984 it would find that buyer in the form of Dresser Industries.

The new Model 30 Haulpaks started to roll off the Peoria production line in May 1963. Rated as a 30-ton capacity hauler, the Model 30 was powered by a 335 gross/310 net horsepower, 6-cylinder Cummins NT-335 Diesel engine, mated to an Allison CLBT-5660 Torqmatic transmission. Overall empty weight of the new model was 46,350 pounds. Image taken in August 1965.

In the summer of 1965 an upgraded Model 30 Haulpak was announced featuring a revised frame with the engine set back further in the chassis, shorting the nose length considerably. Also there were all new upper platform and ladder designs for safer operator use. Along with the Cummins engine, an optional GM 8V-71N diesel, rated at 318 gross/293 net horsepower was now available. A new Allison CLBT-4460 Torqmatic transmission was now standard. Empty weight of this revised model was approximately 47,150 pounds. Shown is one of the "short" nosed Model 30 Haulpaks in December 1966.

The Model LW-32 (formerly the LW-30) was the company's first rear-dump model actually sold to a customer. In August 1958 the LW-32 truck line picked up the Haulpak trade nomenclature, and the hauler's model name changed to that of the Model 32 in early 1959. Standard engine in this early version of the LW-32/Model 32 was a six-cylinder, 375 gross horsepower Cummins NFT-6-BI diesel. At this time the truck would also get a revised front-end design. Picture date is July 1960.

By 1960, the Model 32 Haulpak now featured a 380 gross/355 net horsepower Cummins NT-380, 6-cylinder diesel engine, mated to an Allison CLBT-5840 Torqmatic power-shift transmission. Empty weight was approximately 46,030 pounds. Payload capacity was 32 tons. The Model 32 was replaced with the improved Model 35 in the summer of 1963. Image date is August 1962.

The long running Model 35 Haulpak started officially rolling off the production line in June 1963. Looking much like its predecessor, the Model 32, the new truck featured a slightly larger payload rating of 35 tons. The Model 35 utilized the same Cummins engine and Allison transmission choice as the Model 32, but now offered an optional big 12-cylinder GM 12V-71 diesel, rated at 370 gross horsepower, equipped with a revised torque converter. Empty weight was 48,100 pounds. Image date is June 1965.

The WABCO Model 35 Haulpak would receive a complete facelift in the summer of 1965, including a new shorter front end, revised frame, and new platforms and ladders. Engine choices were the same as the previous design, but the power ratings for the GM 12V-71 were now up to 430 gross/405 net horsepower. Empty weight was now approximately 50,200 pounds. Payload rating remained unchanged at 35 tons. Shown is the first production example of the revised Model 35 Haulpak in June 1965.

WABCO would launch its improved 35-ton capacity Model 35C Haulpak in November 1969. Improvements included a larger standard tire and wheel package of 18.00-33, 24PR size, and more powertrain choices. These included a 6-cylinder Cummins NTA-855-C diesel rated at 420 gross/387 net horsepower, and a 12-cylinder Detroit Diesel 12V-71N rated at 456 gross/420 net horsepower. Transmission of choice was a power-shift Allison CLBT-750 unit for both. Empty weight was approximately 57,110 pounds. The 35C was replaced by the 35D in mid-1980 featuring a new cab design. The 35-series would finally come to an end in 1986 with a proud history of service of approximately 23 years of continuous production. Shown in October 1971 is a 35C alongside the giant 3200, the largest truck in the Haulpak line at the time.

Introduced in the summer of 1962, the WABCO Model 45 Haulpak rear-dump truck was one of the company's newer generation designs featuring a 12-foot, 8-inch wheelbase. Payload capacity was 45 tons, with an empty vehicle weight of approximately 62,400 pounds. Pictured in April 1962 is the pilot Model 45 Haulpak.

The WABCO Model 45 Haulpak was powered by an 8-cylinder, turbocharged Cummins VT-8-430 diesel engine, rated at 430 gross horsepower, mated to a power-shift Allison CLBT-5860 Torqmatic transmission. Standard tires were 18.00-33, 32PR series type. The Model 45 grew out of the Model 42 Haulpak truck program from late 1960. Production records seem to indicate that few (if any) Model 42 trucks were ever actually built. Picture date is April 1962.

The WABCO Model 45 Haulpak would quickly become the Model 50 in February 1963, with the first truck actually produced in June of that year. The Model 50 looked much like the original Model 45 it replaced, but a host of improvements moved the truck design up into the 50-ton payload class range. To handle this extra tonnage, larger 21.00-35, 28PR tires were specified. Wheelbase was the same as the Model 45 at first, but changed to 12 feet, 7 inches in 1965. Empty weight of the Model 50 had also increased to 69,460 pounds. Image taken in May 1965.

The WABCO Model 50 Haulpak offered three powertrain options: a 12-cylinder Cummins V12-525 (525 gross horsepower), a 16-cylinder Detroit Diesel 16V-71 (580 gross horsepower), and a 12-cylinder Detroit Diesel 12V-71 (475 gross horsepower). All engines utilized full power-shift Allison Torqmatic transmissions. Pictured is the Model 50 Haulpak at the 1965 American Mining Congress show held in October of that year.

As the Model 50 matured in the Haulpak line, it would receive a steady diet of improvements and design upgrades. In 1970 three engine packages were offered which included the Cummins VT-1710-C (635 gross/600 net horsepower), the Cummins V-1710-C (500 gross/474 net horsepower), and a Detroit Diesel 16V-71N (635 gross/600 net horsepower), all specified with Allison transmissions. Empty weight had increased to approximately 75,340 pounds. The Model 50 became the Model 50B in March 1980. Production would finally come to an end for the 50-series in 1986. Image date is January 1970.

The LeTourneau-Westinghouse Model 60 Haulpak from 1960 was the company's first large mining truck design to feature a wheelbase of 13 feet, 4 inches. This popular wheelbase would also form the basis for future Haulpak designs, such as the Model 65 and 75 series product lines. Pictured in September 1960 is the pilot Model 60 Haulpak.

The Model 60 looked much like previous Haulpak designs, only proportioned larger. Payload capacity of the new truck was 60 tons, with any empty weight of 66,000 pounds. Tire size was the 21.00-35, 36PR series type. Pictured in September 1960 is the pilot Model 60 shown next to a Model 32. Note the use of the old "LW" logo plate on the nose of the Model 32.

At the time of its introduction, the Model 60 Haulpak was powered by a 12-cylinder Cummins VT-12-700 diesel engine, rated at 550 gross horsepower. A power-shift Allison CLBT-5940 Torqmatic transmission completed the powertrain package. Shown in November 1960 is an early Model 60 working at a Kennecott Copper's Ruth Mine near Ely, Nevada.

The WABCO Model 60 Haulpak would undergo a host of improvements throughout the 1960s and 1970s, including the redesign of the front end to relocate the engine further back in the frame for better weight distribution. New engine choices included the 12-cylinder Cummins VT-1710-C (635 gross/600 net horsepower), and the 16-cylinder Detroit Diesel 16V-71N (608 gross/576 net horsepower), both utilizing Allison CLBT-6061 Torqmatic transmissions. Standard tires were now specified as 24.00-35, 36 PR type. Empty weight was listed at 85,500 pounds. Pictured is a Model 60 in the early 1970s.

The last version of WABCO's popular 60-series Haulpak was the Model 60B, introduced in the summer of 1981. Biggest news for the 60B was its new cab design, plus more power available from its standard engine choices. The Cummins VTA-1710-C was now rated at 675 gross/635 net horsepower, while the Detroit Diesel 16V-71T carried a 682 gross/642 net horsepower rating. Production would come to a close on the 60B in 1986. Spanning some 26 years, the Model 60-series was the company's longest running truck line to carry the Haulpak trade name. Pictured in August 1981 is the first 60B Haulpak produced.

About a year after the introduction of the Model 60, LeTourneau-Wesstinghouse had a new larger hauler ready to join it in the product line—the Model 65 Haulpak. The Model 65 rear-dump was released in the fall of 1961, and like the Model 60, utilized the 13-foot, 4-inch wheelbase and basic frame design. Payload capacity of the new mining truck was rated at 65 tons. Shown at the Peoria manufacturing plant in September 1961 is the prototype Model 65 Haulpak.

The Model 65 was one of the first rear-dump Haulpak designs to be offered with a factory-built coal dump body. Capable of hauling an average 73-cubic-yard heaped (64-cubic-yard struck), 65-ton payload, it proved to be a good alternative method of coal hauling for operations not wanting to employ bottom-dump tractor-trailer units. Pictured is an early Model 65 equipped with one of the company's first coal bodies factory-built for a rear-dump truck in October 1961.

The first Model 65 Haulpaks were equipped with a Cummins VT-12-700 diesel engine (de-rated to 600 gross horsepower). Starting in 1963 a new Detroit Diesel 16V-71 (580 gross horsepower) joined the options list. Transmission of choice was the tried and true power-shift Allison CLBT-6060 Torqmatic. Shown in November 1962 is a special order Model 65 equipped with a high density ore dump body for extreme service use, built for the U.S. Steel Company.

In 1964 the Detroit Diesel engine offered in the 1963 Model 65 was replaced with a more robust 635 gross/600 net horsepower 16V-71N unit. A Cummins VT-12-635 was also offered with the same power output as the big Detroit. Transmissions were also updated for both the Cummins and Detroit Diesel offerings to the Allison CLBT-6061 Torqmatic. Standard tires of choice were 21.00-35, 36PR series type. Approximate empty vehicle weight was listed at 76,000 pounds. In December 1965 the Model 65 became the 65A (which was based on the new 75A Haulpak, also introduced at the same time), and then in 1967 this model was replaced by the 65B (which, again, was primarily based on the upgraded Model 75B platform), equipped with a slightly less powerful engine. The 65B nomenclature would disappear in 1969, with the model simply becoming a series of options under the Model 75B truck line. Pictured is a Model 65 Haulpak at a mining operation run by Sherman Mines in Ontario, Canada, in June 1966.

WABCO's first endeavor at producing a diesel electric drive mining truck was a special Model 65 Electric Drive Haulpak, sometimes referred to as the 65E. The 65E utilized electric drive components supplied by General Electric (GE), and was powered by either a 12-cylinder Cummins VT-12-700, or a 16-cylinder Detroit Diesel 16V-71T unit. Both engines were rated at 700 gross/625 net horsepower each. Braking systems were a combination of a standard 900-horsepower electric dynamic brake (1,200-horsepower optional) and mechanical shoe type service brakes. Payload capacity was 65 tons, with an empty weight of approximately 85,000 pounds. Shown in August 1962 is the pilot Model 65 Electric Drive Haulpak.

Unlike the GE drive system being experimented with by Unit Rig in 1962, the WABCO/LeTourneau-Westinghouse design did not use an electric motorized wheel design. Instead, the Model 65E was designed around a GE GT-603-D generator and the 769-BI electric drive motor. This motor drove a final drive, power transfer, no slip rear differential. Pictured is the original layout of the upper deck of the 65E showcasing the electric dynamic resistance grid and system control cabinets. Image date is August 1962.

The prototype WABCO Model 65 Electric Drive Haulpak was tested throughout 1963 at Anaconda's Berkeley Pit in Butte, Montana. Engineering experience gained from these tests led the company to build a second prototype Model 65E in 1964. This truck featured a redesigned GE dynamic resistance grid and main generator. This model was also powered by the Detroit Diesel 16V-71T engine. The Model 65E was limited in availability, but would eventually lead the company to its first true, full production diesel electric drive rear-dump, the Model 120 Haulpak. Shown in June 1964 is the revised Model 65 Electric Drive Haulpak.

The largest Haulpak truck to utilize the company's 13-foot, 4-inch wheelbase chassis was the Model 75A. Introduced in 1965, the 75A was rated as a 75-ton capacity hauler in rear-dump form. Powertrain options were a Cummins VT12-700, or a Detroit Diesel 16V-71NT, rated at 700 gross/665 net horsepower each. The same power-shift Allison DP-8860 Torqmatic transmission was specified for both engine choices. Standard tire size was 21.00-35, 40 PR type. Empty weight was listed at 89,400 pounds. Pictured in September 1965 is the first Model 75A Haulpak built.

The WABCO Model 75A would become the 75B in mid-1967. The 75B featured many design improvements, which included new front wheel fabrications. Standard engines were listed as the Detroit Diesel 16V-71T, and the Cummins VTA-1710-C. Both engine power ratings were the same as the figures for the Model 75A. Standard transmission choice was the Allison DP-8961 Torqmatic. Tires were now larger 24.00-35, 42PR (E-3) series type. Empty weight of the rear-dump version was 91,500 pounds. The first Model 75B Haulpak built in May 1967 was not a rear-dump version, but a tractor unit for pulling a coal bottom-dump trailer. Pictured in October 1973 is a full production Model 75B.

The last version of the 75-series Haulpaks produced by WABCO was the Model 75C in 1981. The Model 75C featured many improvements that increased the truck's overall performance and reliability in the field, such as an all new operator's cab design. Engines were the same as those specified for the 75B, but power had increased to 725 gross/679 net horsepower for both powerplants. General specifications such as tires and weights remained unchanged from the previous model. The 75C would remain in the product line until 1986. Shown is the prototype Model 75C Haulpak in June 1981.

WABCO originally announced its 85-ton capacity Model 85C Haulpak rear-dump truck in October 1970. But it would not be until the summer of 1971 that the first unit would roll off the Peoria assembly line. The 85C was built on a chassis layout based on a 14-foot, 7-inch wheelbase, with features patterned after the frame design utilized on their 120B electric-drive mining truck. Pictured in June 1971 is the pilot 85C equipped with the standard rock dump body that was offered in 60- to 70-cubic-yard sizes.

The production WABCO Model 85C Haulpak was offered with two standard and two optional engine choices. These included a Detroit Diesel 16V-71T (700 gross/665 net horsepower), two Cummins VTA-1710-C units (standard 700 gross/665 net horsepower, and optional 800 gross/765 net horsepower), and a Caterpillar D348 diesel (850 gross/815 net horsepower). All utilized the Allison DP-8961 Torqmatic transmission with electric shift control. The 85C shown is equipped with a factory-built high volume coal body, available in sizes from 94 to 108 cubic-yards. Payload remained fixed at 85 tons in capacity. Image date is July 1971.

For extreme hauling duties, WABCO offered the Model 85C Haulpak with an optional taconite ore dump body. Capacity of these ore bodies ranged from 47 to 59 cubic-yards, and proved very popular in the marketplace. Specified standard tires for the 85C were 24.00-49, 42PR (E-3) series type. Average empty weight for the model was 106,260 pounds (rock body). Pictured in September 1971 is the first ore body built for the 85C.

The last version of the company's popular 85-ton capacity truck to be placed into service was the Model 85D Haulpak from 1980. The updated model now featured a new operator's cab design, and simplified engine choices. Standard powerplant offerings were the 12-cylinder Cummins KT-2300-C, rated at 900 gross/858 net horsepower, and the 16-cylinder Detroit Diesel DDAD 16V-92T, rated at 860 gross/818 net horsepower. All utilized the Allison electric shift DP-8961 Torqmatic transmission. The first 85D was completed in April 1980. The truck shown was photographed one year later in April 1981.

The WABCO Model 85D Haulpak persevered for the company even in the difficult economic recessionary times of the early 1980s in the world marketplaces. Standard tire size was the same as the 85C, but a larger optional 27.00-49, 36PR (E-3) mounted on bigger rims was now offered. Empty weight of the 85D was approximately 119,300 pounds. In late 1986 the WABCO Model 85D was renamed the Dresser 325M. Image taken in November 1981.

One of WABCO's larger mechanical-drive mining truck offerings was the Model 100 Haulpak. Officially unveiled at the October 1982 AMC show in Las Vegas, Nevada, the Model 100 would be displayed alongside its larger mechanical-drive brother, the 140CM Haulpak. The Model 100 was powered by either a Cummins KTA-2300-C diesel engine, rated at 1,050 gross/962 net horsepower, or a Detroit Diesel 12V-149T, rated at 1,050 gross/952 net horsepower. The only transmission of choice was the Allison automatic electric shift DP-8962 Torqmatic. Wheelbase of the new design was 16 feet, 3 inches, with a standard tire size of 30.00-51, 40PR. Image taken in April 1984.

The WABCO Model 100 Haulpak was capable of hauling a 100-ton payload, and had a listed empty vehicle weight of 160,600 pounds. The company's new 100-ton capacity truck was slow to be accepted in the marketplace, due mainly to the worldwide economic recession going on at the time. Mining operations relied on tried and proven designs during these troubled times and shied away from untested new model introductions. In late 1986 the WABCO Model 100 was reclassified as the Dresser 385M Haulpak. Image date is April 1984.

It is probably an understatement to say just how important the Model 120A electric wheel drive truck program was to WABCO. This model would lay the groundwork for all future electric-drive Haulpak designs to be built, not only by WABCO, but Dresser and Komatsu. Pictured at the company's Peoria proving grounds in June 1965 is the pilot Model 120A Haulpak.

The Model 120A was not the first electric-drive Haulpak (Model 65E), but it was the first for the company to utilize the General Electric motorized wheel system. The heart of the DC electric-drive system was a GE GT-603-J generator, and a pair of GE 772 wheel motors, mounted in the rear wheel housings. The early Model 120A was powered by a big 2,392-cubic-inch, 12-cylinder Fairbanks Morse (Dorman) 50A6TCW-12V engine, rated at 930 gross/845 net horsepower. The hauler also featured a new operator's cab design as well. The 120A was officially unveiled by WABCO at the October 1965 AMC show in Las Vegas, Nevada. Image taken in June 1966.

The payload capacity of the original Model 120A Haulpak was rated at 105 tons. This was carried on standard 27.00-49, 48PR tires. Wheelbase of the truck was 15 feet, 10 inches. Empty weight of the hauler was listed at 155,800 pounds in standard trim. Shown is a 105-ton capacity Model 120A operating at Kennecott Copper's Ruth Mine located just outside of Ely, Nevada, in September 1966.

In 1967, WABCO started offering the Model 120A equipped with a new 12-cylinder Detroit Diesel 12V-149T engine (first tested in a 120A chassis in June 1966), which was far superior to the original Fairbanks Morse unit. Power on the new Detroit was up to a healthy 1,000 gross/920 net horsepower. Units that had shipped with the original Fairbanks engines were eventually upgraded in the field with the new Detroit powerplant. With more power came an increase in payload, now job rated at 110 to 120 tons. In mid-1968 WABCO offered an electric-drive Model 100A Haulpak rated at 100 tons in capacity. This truck was basically a 120A fitted with a less powerful engine and a smaller dump body. In September 1969 WABCO announced the availability of an improved Model 120B Haulpak. The 120B featured the same Detroit Diesel found in the last version of the 120A released, but added a 12-cylinder Caterpillar D348 diesel (990 gross/910 net horsepower) to the options list. Payload capacity range was listed at 110 to 130 tons, depending on tire choices. For 130 tons, the 30.00-51, 52PR tire size was required. Empty vehicle weight was listed at 174,300 pounds. Wheelbase had also been increased by one foot to 16 feet, 10 inches. Pictured in January 1970 is the prototype Model 120B Haulpak.

The WABCO Model 120-series proved to be a great success in the marketplace, and was the perfect candidate for big overseas sales as well. Shown at the Peoria plant in June 1972 is an export 120B ready to be shipped to Russia. The truck is equipped with a special order Cummins KTA-2300-C diesel engine, and an extreme-weather package to help it cope with the conditions it will encounter in Russia.

In early 1975 WABCO released information on a new Model 120C Haulpak, which would begin shipping in July of that year. Featuring numerous performance and reliability upgrades, the 120C offered no less than six engine choices. These included the standard Detroit Diesel 12V-149T (1,000 gross/895 net horsepower), and optional Caterpillar D348 (990 gross/866 net horsepower) and D349 (1,130 gross/1,000 net horsepower), Cummins KTA-2300-C1050 (1,050 gross/950 net horsepower) and KTA-2300-C1200 (1,200 gross/1,075 net horsepower) powerplants, plus an additional Detroit Diesel 12V-149TI (1,200 gross/1,070 net horsepower). For the bigger 1,200 horsepower rated engines, larger GE 776H electric wheel motors were recommended. Payload capacity was rated at 120 tons. Shown in March 1982 is a 120C being loaded at a Hawkeye Coal Co. mine located in Pikeville, Kentucky.

The success in the marketplace of the 120A/B/C Haulpaks led WABCO to introduce a new mechanical-drive version of the truck based on the proven 16-foot, 10-inch wheelbase and chassis layout—the Model 120CM. The first 120CM was officially unveiled at the October 1978 AMC show in Las Vegas, Nevada. The big mechanical-drive Haulpak could be ordered with a Cummins KTA-2300-C (1,200 gross/1,075 net horsepower) or a Detroit Diesel 12V-149TI (1,200gross/1,070 net horsepower) engine. Transmission of choice was a fully automatic WABCO MTC-93 mechanical drive unit manufactured for WABCO by Twin Disc, Inc. Payload capacity was 120 tons, carried on standard 30.00-51, 46PR, or optional larger 33.00-51, 50PR tires. Empty weight was approximately 177,830 pounds. Image date is September 1979.

In November 1981, WABCO released information on the next version of its popular 120-ton capacity electric drive Haulpak, the Model 120D. Standard engine choices were now a Cummins KTA-2300-C-1200 (1,200 gross/1,072 net horsepower), or a Detroit Diesel 12V-149TI (1,200 gross/1,050 net horsepower). Two optional 1,050 gross horsepower powerplants could also be requested by the customer. The GE 772YS type wheel motors were offered as standard, with more powerful 776HS units required for the special heavy-duty 120D version equipped with the deep pit electric system. In late 1986 the WABCO 120D became the Dresser 445E Haulapk in the product line. Empty weight of the "D" model was 178,235 pounds. Shown at work in November 1983 is a Model 120D at Line Creek Coal, located near Sparwood, British Columbia, Canada.

In a quest for more horsepower in a lighter weight powertrain package for superior power to weight ratios, a few of the heavy-equipment manufacturers saw the benefits that turbine powered mining trucks might offer. WABCO's entry into this grand experiment was the Model 120A Turbine Haulpak. The 120A turbine program started in 1967 with the modification of the Model 160A rear-dump tractor-trailer prototype. This unit had its original Fairbanks Morse engine removed and replaced with a lightweight but powerful IH Solar Turbine unit. Engineering knowledge gained from this test program led directly to the 120A Turbine. Pictured in April 1969 in Peoria is one of the first 120A Turbine Haulpak chassis built.

WABCO built a total of five Model 120A Turbine Haulpaks. Four units were shipped to Kennecott Copper's Bingham Canyon Mine, located just outside Salt Lake City, Utah (and were up and running by August 1969), and one truck was shipped to Cleveland Cliffs-Marquette Iron Mining Company's Republic Mine, located in the Upper Peninsula of Michigan. Payload capacity range of the turbine trucks was 110 to 120 tons. Shown in September 1969 is the lone 120A Turbine shipped to the Republic Mine.

The WABCO Model 120A Turbine Haulpaks had the power engineers were hoping for, but the fuel consumption of the 1,100 horsepower Solar Turbine engine was enormous by anyone's standards. Other drawbacks were how loud the engines were in service, making life rather difficult for the people that operated and serviced them. Unfortunately, the fuel consumption of the 120A Turbines (and their initial cost of the engines) over the long term cancelled out the power benefits the engines provided. Image taken at the Republic Mine in September 1969.

The largest of WABCO's mechanical-drive Haulpak designs was the Model 140DM from late 1982 (briefly referred to as the 140CM). The heart of the 140DM was its mechanical-drive powertrain layout. Power choices included a Cummins KTA-2300-C, or a Detroit Diesel 12V-149TI, rated at 1,350 gross/1,242 net horsepower each. The transmission consisted of an 8-forward speed, fully automatic WABCO MTC-93 mechanical drive unit manufactured by Twin Disc, Inc. The standard tire choice was 33.00-51, 50PR (E-4) series type, utilized on a chassis wheelbase of 16 feet, 10 inches. Maximum payload capacity was listed at 140 tons, with an empty vehicle weight of approximately 200,500 pounds. The 140CM/DM was officially unveiled at the 1982 AMC show in Las Vegas. The WABCO Model 140DM would become the Dresser 510M in late 1986. Image taken in October 1982.

Originally announced by WABCO in September 1968, the 150-ton capacity Model 150B Haulpak had a back-order of more than 30 trucks even before the first unit was completed in 1970. Built on the same 16-foot, 10-inch wheelbase chassis layout as the Model 120B, the 150B featured more power, beefier electric drive components, and a larger payload capacity. Standard engine choice was the Detroit Diesel 16V-149T (1,325 gross/1,185 net horsepower), with optional 1,600 gross horsepower Detroit Diesel 16V-149TI and 1,200 gross horsepower Caterpillar D349 units available. The main electric drive system was comprised of a powerful GE GTA-15 alternator and a pair of 776 wheel motors. Image date circa 1972.

In late 1972 WABCO offered a special version of its 150-ton capacity Haulpak in the form of the Model 150Bw. The 150Bw shared most of the same technical specifications as the 150B, except when it came to the electric drive system. WABCO engineers replaced the General Electric components with a new Westinghouse Electric Transmission package, which included the alternator, and a single electric drive motor with a WABCO planetary drive axle. Empty vehicle weight of both models of the 150-ton payload Haulpak was approximately 220,200 pounds. Standard tire size for both models were 36.00-51, 42PR (E-4) series type. Pictured in December 1972 at Anamax's (Anaconda/AMAX) Twin Buttes Mine located in Sahuarita, Arizona, is one of the first Model 150 Bw Haulpaks produced.

The WABCO Model 150B would become the 150C in early 1974. Later in the year an improved 150Cw was also introduced. The 150C/Cw Haulpak now listed the powerful Detroit Diesel 16V-149TI (1,600 gross/1,440 net horsepower) as the engine of choice for both model types. Capacity remained unchanged at 150 tons, but empty vehicle weight had increased to 212,500 pounds. The 150C utilized a General Electric drive system, with a Westinghouse Electric drive setup installed in the 150Cw. Wheelbase was listed at 17 feet, 10 inches. Shown in September 1974 is the first 150Cw chassis to roll off the Peoria assembly line.

The Model 150C was only in the product line for a few months before WABCO replaced it with the larger capacity 170-ton capacity Model 170C Haulpak in mid-1974. The standard engine available was the Detroit Diesel 16V-149TI. In 1978 a Cummins KTA-3067-C was added to the options list. Powerplants were rated at 1,600 gross/1,450 net horsepower each. The electric drive system was all GE, with powerful 776 series motors mounted in the rear wheel assemblies. Wheelbase was the same as the 150C. Empty weight of the 170C was listed at 216,700 pounds. Standard tire choice was the 36.00-51, 50PR (E-4) type. In April 1975 a Model 170Cw, equipped with the Westinghouse Electric transmission package, replaced the previous 150Cw offering. Pictured in June 1974 is the first 170C Haulpak produced.

The Model 170C Haulpak proved to be a real performer for WABCO in the marketplace, especially in mining operations outside of North America. In mid-1980 the company released an upgraded "D" version of its winning 170-ton capacity truck. The Model 170D Haulpak utilized the same engine choices as the 170C offered in 1978, but net horsepower was now up to 1,470 for both the Detroit Diesel and Cummins powerplants. An improved GE GTA-22 alternator and 776S wheel motors completed the powertrain package. Empty vehicle weight had increased a bit to 223,400 pounds. There was no upgrade for the 170Cw because of its removal from the product line in 1978. The WABCO 170D was renamed the Dresser 630E in late 1986. Shown in September 1980 is the pilot Model 170D.

The last new truck program to launch before the WABCO name disappeared from the Haulpak truck line was the Model 190. Released in mid-1985, the 190 Haulpak utilized the same 17-foot, 10-inch wheelbase chassis layout as the 170D. Engines specified for the 190-ton hauler were a Detroit Diesel 16V-149TIB, or a Cummins KTTA-50-C, both rated at 1,800 gross/1,704 net horsepower. Standard specified tire size was 36.00-51, 58PR type. Pictured in June 1985 is the prototype Model 190 (minus dump body), equipped with factory yard tires which were used around the Peoria plant only.

The WABCO 190 Haulpak utilized a diesel-electric drivetrain utilizing a GE GTA-22 alternator and two powerful 788 wheel motors. Maximum payload capacity was 190 tons, with an empty vehicle weight of approximately 250,000 pounds. The Model 190 did not have a direct counterpart in the new Dresser Haulpak lineup, but instead became a series of upgrade options offered for the 630E truck platform. Only a handful of WABCO 190 Haulpaks were delivered before the Dresser product name change in 1986. Shown is a WABCO 190 working at the BHP LaPlata Coal Mine, located north of Farmington, New Mexico, in August 1986.

WABCO's largest rigid-frame, rear-dump Haulpak design was its famous Model 3200. The diesel-electric drive Model 3200 was the company's only three-axle, tandem-drive axle design built around a rigid-frame layout. The company originally released concepts of the truck at the fall 1969 AMC show. The first completed unit rolled out of the Peoria assembly plant in September 1971. Image taken in early October 1971.

The WABCO Model 3200 Haulpak was designed around its massive powertrain provided by GM's Electro-Motive Division (EMD). The big three-axle hauler was powered by a 12-cylinder, 2-stroke, EMD 12-645-E4 locomotive diesel engine. In its original form, the engine was rated at 2,000 gross/1,800 net horsepower. The rear drive wheels were driven by two EMD type D79X3A electric traction motors, with one motor mounted externally to the center case structure of each rear axle assembly. These provided power to each drive axle through a gear coupling. The 3200 did not utilize the use of electric wheel traction motors in its rear wheel hub assemblies like other Haulpak two-axle designs of the time. Image date is October 1971.

Initially, the WABCO Model 3200 payload rating was only 200 tons in capacity for the early prototype units. But by 1973 payload capacity had been raised to 235 tons with the use of 36.00-51, 42PR (E-4) tires. The original 200-ton capacity units were equipped with 33.00-51, 42PR (E-4) series type. Empty vehicle weight of the big Haulpak was 327,550 pounds with a full fuel load. Image taken in October 1971.

WABCO shipped its first two prototype Model 3200 Haulpaks in 1972. The first unit shipped in the summer of 1972 to the Mt. Newman Mining Co. Pty., Ltd. operation at Mt. Whaleback in Western Australia. In the fall of 1972 truck number two started testing at the Pima Mine located south of Tuscan, Arizona. Pictured in late 1972 is the Pima truck in operation.

The largest tires available for the 3200 Haulpak in its initial design stage were the 33.00-51, 42PR (E-4) series type. This, more than anything else, limited the early trucks to 200 tons in capacity. Once the optional 36.00-51, 42 PR size became available, payload was increased to 235 tons. The tandem-drive layout of the truck made it a very robust design. But care going down steep grades in the wet were required because of the trucks extreme tendency to understeer, that is to continue to go straight even though the operator is steering to the left or right. Image taken at the Pima Mine in late 1972.

Shown in March 1973 is the number two 3200 Haulpak with a full 200-ton-plus payload coming out of the main pit at the Cyprus' Pima Mine. On level ground with a full load on its back, the Model 3200 was capable of attaining a leisurely 25-miles-per-hour top speed.

In mid-1974 WABCO announced the availability of an upgraded "B" version of its huge Model 3200 Haulpak. The Model 3200B featured many improvements that helped in the reliability and performance of the hauler in the field, including frame upgrades, and the use of optional 36.00-51, 50PR (E-4) tires. At this time in the 3200B Haulpaks life, its engine output was still rated at 2,000 gross horsepower. Image taken in Peoria in August 1976.

In 1976 the WABCO Model 3200B was refined even further with a power increase from its big EMD diesel to 2,475 gross/2,250 net horsepower. With the power bump the larger, higher ply 36.00-51, 50PR (E-4) tires became standard. These two factors, more than anything else, raised the maximum payload capacity of the 3200B to 250 tons (260 tons when ordered without bed liners). When equipped with the 42-ply rated tires, payload dropped to 235 tons. Pictured in August 1976 is the higher horsepower 3200B equipped with the new larger tires.

The heart of the 3200/3200B was its huge GM Electro-Motive Division (EMD) 12-645-E4 diesel engine. The locomotive engine designed and built by EMD was as bullet-proof of a design as one would want in a haul truck. But the weight penalty for such a heavy-duty, low-revving diesel was substantial. Weight of the 7,740-cubic-inch (126.9 liter) 12-cylinder, 2-cycle engine alone was 29,000 pounds. The 12-645-E4 powerplant was rated by EMD at 2,475 gross horsepower. But WABCO chose to de-rate the engine down to 2,000 gross horsepower to guarantee its reliability in the field. Once tires of suitable size were made available to the 3200B, the power was ratcheted back up to a full 2,475 gross horsepower (2,250 net) at 900 rpm. A 2,000 gross horsepower version was still offered for Australian market 3200B Haulpaks. Image date is November 1974.

The 3200B was one of the best looking Haulpaks WABCO engineers ever conceived. But various industry declines, first in the early 1970s in Australia, and then the worldwide recession of the early 1980s, crippled it in the marketplace. As time passed, engines and tire sizes in the industry had increased enough to make similar-sized trucks to the 3200B with a two-axle design, making the three-axle layout costly and unnecessary. Shown is a 3200B Haulpak at Duval Corporation's Sierrita Mine, located south of Tucson, Arizona, in June 1976.

As the WABCO 3200/3200B increased in payload size, the length of the truck model also grew. Length of the trucks was 50 feet, 6 inches for 200 tons in capacity; 53 feet for 235 tons; and 54 feet, 3 inches for 260 tons. Overall width of the 3200B was 25 feet, 2 inches. Empty vehicle weight of the "B" model was 365,000 pounds. Shown in October 1975 is a 3200B operating at Lornex Mining Corporation's Logan Lake Mine in British Columbia, Canada.

One of the only WABCO 3200B Haulpaks to be painted in the company's revised white color scheme was the show truck on display at the October 1978 AMC show in Las Vegas. The 3200-series operated in mining operations in the United States (Cyprus' Pima Mine, Duval's Sierrita Mine, and Kennecott Copper's Ray Mine), Canada (Lornex Mining's Logan Lake Mine), and Australia (Mt. Newman Mining's Mt. Whaleback, and Hamersley Iron's Mt. Tom Price). By the end of 1978, 40 trucks were listed as being in service worldwide; 18 in North America, and 22 in Australia (the WABCO Rydalmere plant was responsible for the production of 10 out of the 22 units sold in Australia). Today, at least two 3200-series Haulpaks survive and are on public display. One unit is in Newman, in Western Australia; the other one is in Logan Lake, British Columbia.

One of the more unique Haulpak concepts to be placed into limited service was the electric four-wheel-drive, tractor-trailer rear-dump hauler program. The pilot version of this hauler was the Model 160 Haulpak. The Model 160 consisted of a 120A-series tractor unit and a rear-dump trailer unit. In this configuration, the rear wheels of the tractor and trailer were equipped with electric wheel traction motors, giving the model four-wheel drive ability. Shown in October 1965 in Peoria is the prototype Model 160 Haulpak as it is being pulled outside in completed form for the first time, minus the trailer's outside wheels.

The prototype WABCO Model 160 Haulpak was powered by a 12-cylinder, 4-stroke Fairbanks Morse (Dorman) engine, rated at 930 gross/845 net horsepower. The General Electric drive system consisted of a GE GT-603-J generator, plus four 772-C2 electric traction wheel motors, mounted in the rear wheel assemblies of the 120A tractor and the rear-dump trailer. Overall length of the rig was 52 feet, 10 inches in length, with a maximum width of 20 feet, 8 inches. Standard tire size was 27.00-49, 42PR series type. Image taken in November 1965.

The Model 160 Haulpak was capable of hauling a 160-ton payload in its 119-cubic-yard rear-dump body. Empty vehicle weight was approximately 199,000 pounds (tractor at 103,000 pounds/trailer at 96,000 pounds). In mid-1967, the original Fairbanks Morse 12-cylinder engine was removed from the 120A tractor chassis of the prototype Model 160 and replaced with a 1,100 gross horsepower IH Solar Turbine powerplant. The turbine-engined 160 Haulpak was then shipped to Kaiser Steel's Eagle Mountain Mine in California for further testing of the experimental drivetrain (in full service by October 1967). The modified Model 160 essentially became the test-mule for the two-axle 120A Turbine rear-dump truck from 1969. Pictured is the Fairbanks Morse equipped Model 160 at WABCO's Peoria proving grounds in March 1966.

WABCO continued on with the development of its electric-drive, tractor-trailer concept with the engineering knowledge gained from the Model 160 program. In 1969 the company completed its first Model 200 Haulpak consisting of a tractor based on the 16-foot, 10-inch wheelbase layout of the new 120B/150B chassis, and a new 200-ton capacity rear-dump unit. In late 1970, the model was reclassified as the 200B. The 200B Haulpak was powered by a standard 16-cylinder Detroit Diesel 16V-149T engine, rated at 1,325 gross/1,230 net horsepower. An optional higher powered 16V-149TI (1,600 gross/1,440 net horsepower) was also available. The electric drive system was all General Electric, with a GE GTA-15 alternator and 772-K wheel motors. Standard tire size on this new model was 30.00-51, 46PR series type. Pictured in July 1969 is the prototype Model 200 (200B) at the main Peoria plant.

WABCO delivered a small fleet of Model 200B Haulpaks to Duval Copper's Sierrita Mine, located south of Tucson, Arizona, with the first unit up and running by October 1969. In 1971 WABCO engineers made quite a few design upgrades to the Model 200B, including a redesign of the front of the tractor unit (which now resembled the 150B), relocation of various electrical components, and structural improvements to the rear-dump trailer unit. Pictured is the improved 200B Haulpak in October 1971.

The WABCO Model 200B Haulpak was a very compact and aggressive looking hauler for its day. Overall length of the tractor-trailer unit was 55 feet, 1 inch, with a width of 19 feet, 8 inches. Shown in October 1971 is the redesigned 200B at U.S. Steel's Mountain Iron mining operation in Minnesota, getting ready for its field evaluation testing to begin. At this time, the truck was still owned by WABCO.

The 200B Haulpak payload rating for its 124.3-cubic-yard rear-dump body was 200 tons. The empty vehicle weight was listed at 275,800 pounds (tractor at 137,100 pounds/trailer at 138,700 pounds). The design layout of the 200B, featuring powered drive wheels for the tractor and trailer units, seemed to offer great performance potential. But the company's own new three-axle design, the 3200, seemed to make the 200B a moot point. Engineering development would stop on the 200B program after this redesigned unit, with all focus for a 200-plus-ton Haulpak truck program being shifted to the Model 3200. Image date is October 1971.

One of the first Peoria-designed LeTourneau-Westinghouse trucks to be placed into service was not a rear-dump, but was actually a tractor-trailer coal bottom-dump, the Model LW-75. The original LW-75 was completed in November 1957, and delivered to Midland Electric's coal mining operation in Farmington, Illinois, in December to start a series of demonstrations for Midland mining officials. Midland officials liked what they saw and decided to purchase the pilot bottom-dump, and additional units scheduled for delivery in the second half of 1958. The pilot model carried the identification of the LW-75, but was not yet referred to as a "Haulpak" because the trade name had yet to be adopted for the new truck lines. But units delivered after this would be identified as Model LW-80 "Haulpaks." Only the pilot hauler ever carried the LW-75 model plates. Shown undergoing performance tests at Midland Electric on February 19, 1958, is the pilot LW-75.

In early 1959 the LeTourneau-Westinghouse LW-80 Haulpak identification was changed to simply the Model 80. But trucks would continue to ship with the old LW-80 name plates on them until the existing factory parts supply was exhausted. The production Model 80 was powered by a 12-cylinder Cummins NVH-12-BI diesel engine, rated at 450 gross horsepower. The only transmission of choice was the power-shift Allison CLT-5840 Torqmatic. Shown in September 1959 is a Model 80 Haulpak carrying the "SOUTHWESTERN" mining company name. Southwestern Illinois Coal Corp. was the second fleet buyer of Model 80 Haulpaks (for their Streamline Mine). Their first unit was originally completed by LeTourneau-Westinghouse in November 1958.

The early Haulpak bottom-dump coal haulers were well received in the marketplace, especially in coal mining operations in Illinois, Indiana, and Kentucky. The Model 80 was capable of hauling an 80-ton payload of coal in its 100-cubic-yard struck trailer. The trailer unit was a box-beam-and-corrugated steel construction design, with air-actuated bottom-dump doors. A Hydrair suspension system was utilized on all wheels and the main trailer hitch. Overall length of the Model 80 was 55 feet, 9 inches, with an empty weight of 73,000 pounds. Standard tire size was 18.00-33, 32PR series type. Shown in December 1959 is a Model 80 ready for delivery to Peabody Coal Company's Ken Mine.

LeTourneau-Westinghouse first announced the availability of a larger 90-ton capacity coal hauler in November 1959 in the form of the Model 90 Haulpak bottom-dump. But it would not be until December 1960 that the first unit was ready for shipping to the customer's mine site, in this case Peabody Coal Company. Once production finally got underway, the Model 90 would replace the Model 80 in the Haulpak product line. Pictured in June 1962 is a Model 90 Haulpak being loaded at Enos Coal, located near Arthur, Indiana.

The Model 90 Haulpak was powered by either a 12-cylinder Cummins NVH-12-550 engine (550 gross horsepower) or a 16-cylinder Detroit Diesel 16V-71 (580 gross horsepower). Standard transmission was a power-shift Allison CLT-5940 Torqmatic, with the availability of an optional LeTourneau-Westinghouse Powerflow "700" unit. Standard tire choice for the tractor was 18.00-33, 32PR, while the trailer had the slightly larger 18.00-49, 32PR series type. Shown in July 1962 is a Model 90 in service at Tecumseh Coal in Booneville, Indiana.

The payload capacity for the Model 90 was a nominal 90 tons, with a maximum capacity of 100 tons, with bottom-dump trailers capable of hauling 100 or 112 cubic-yards of coal. Approximate empty weight of the unit was 77,400 pounds. Overall length of the Model 90 was 56 feet, 1 inch. Pictured in January 1964 is an Enos Coal Company bound WABCO Model 90 Haulpak featuring the 112-cubic-yard coal hopper equipped with Rockwell shoe brakes.

The next step up in size for WABCO coal haulers came in early 1964 with the release of the Model 120 Haulpak. The Model 120 consisted of a modified Model 65 tractor chassis and a WABCO-designed 120-ton capacity bottom-dump trailer. Engines of choice were the Detroit Diesel 16V-71N, or the Cummins VT-12-635, both rated at 635 gross/600 net horsepower. Transmission of choice was a power-shift Allison CLBT-6061 Torqmatic. Tires on the tractor were 21.00-35, 36PR series, with the trailer fitted with larger 21.00-49, 36PR type. Shown in Peoria in February 1964 is the first Model 120 Haulpak produced. Built for the Peabody Coal Company, it would be shipped to their Sinclair Mine in Kentucky.

Starting in mid-1966 new versions of coal and dirt bottom-dump models would now simply become options for a particular Haulpak truck in the product line. One of the first coal haulers to fall under this new identification was the WABCO Haulpak 75B Series Tractor with 120 Bottom Dump Coal Hauler. Power for this 120-ton capacity coal hauler was provided by a Detroit Diesel 16V-71NT engine, or a Cummins VT-12-700 unit, both rated at 700 gross/665 net horsepower. Transmission of choice was the Allison DP-8860 Torqmatic. Tires for the tractor were 24.00-35, 36PR, with the trailer getting 21.00-49, 36PR series. Length of the entire unit was 65 feet, 2 inches, with an empty vehicle weight of 128,800 pounds. In 1974 this unit was referred to as the Model 75B/120CT. By late 1975 it was shortened to simply the Model 120CT. Shown in May 1967 is the first 75B/120 Coal Hauler built ready for delivery to Peabody Coal Company's Sinclair Mine in Kentucky.

WABCO produced most of its Haulpak tractor units, such as this 75B series from May 1973, for use with trailers designed and built by the company. In 1973 WABCO offered two special tractor units for use with Athey bottom-dump wagons. The Model 35CT was meant to go with the 72-ton capacity Athey W4272B Bottom Dump, while the larger Model 50T was matched to the 100-ton capacity Athey W60100B unit. Both were dirt bottom-dump trailers. This was not the first time WABCO offered the Model 50 in tractor form. Back in March 1967 the company built a special order of 120-ton capacity 50/120 Coal Haulers for Southwestern Illinois Coal Corp., for use at its Captain Mine, located near Percy, Illinois.

WABCO had envisioned a 150-ton capacity Haulpak coal hauler as far back as April 1972, when concept artwork was first produced for marketing purposes. By October 1974 this design was now referred to as the Model 85C/150CT Coal Hauler. By mid-1975 the product line name became the Model 150CT. The original 150CT utilized a Model 85C tractor chassis powered by a standard Detroit Diesel 12V-149T engine (1,000 gross/940 net horsepower), with optional Cummins VTA-1710-C (800 gross/765 net horsepower) and Caterpillar D348 (850 gross/815 net horsepower) diesels available, all matched to an Allison DP-8961 Torqmatic transmission. Shown in August 1975 is one of the first Model 150CT Haulpaks produced, soon to be delivered to Decker Coal Company in Montana.

The WABCO Model 150CT would be given an 85D tractor upgrade in mid-1980. This version of the 150CT was offered with either a Cummins KTA-2300-C, or Detroit Diesel 12V-149T engine, both rated at 1,050 gross/967 net horsepower. The only transmission of choice was the Allison DP-8962 Torqmatic. Standard tire size was 27.00-49, 36PR series for the entire unit. This was the same tire size as the previous 150CT version. Capacity of the coal hopper was 150 tons (194 cubic-yards heaped). Overall length was 76 feet, 5 inches, with an empty vehicle weight of 203,500 pounds. The Model 150CT (as well as the 120CT) were both removed from the product line in 1986. Pictured in May 1982 is an upgraded 150CT Haulpak operating at Arco's Coal Creek Mine, near Gillette, Wyoming.

The WABCO Haulpak 75A Series Tractor with 120 Bottom Dump Rock and Dirt Hauler was a custom-built product produced for use on the Blue River Dam project in Oregon. The tractor unit could be specified with a Cummins VT-12-700, or Detroit Diesel 16V-71NT engines, both rated at 700 gross/665 net horsepower, with most of the units shipped to the dam project equipped with the Cummins powerplant. Standard transmission was the power-shift Allison DP-8860 Torqmatic. Pictured in June 1966 is the first 75A/120 Rock and Dirt Hauler built by the factory.

Most of the WABCO Haulpak 75A/120 Rock and Dirt Haulers were shipped to Blue River Construction Company in Oregon, for work on the Blue River Dam project. The heavy-duty bottom-dump trailer was capable of hauling a 120-ton payload in its 86-cubic-yard heaped hopper, and was designed to withstand the loading impact of blasted rock. Standard tires specified for the unit were 21.00-35, 36PR type on the tractor, and much larger 27.00-49, 36PR on the trailer itself. Empty weight of the rig was 153,540 pounds, with an overall length of 63 feet, 11 inches. In 1970 this model was offered equipped with an upgraded 75B Series Tractor unit. Image date is June 1966.

In the 1970s a few heavy-equipment manufacturers fell in love with the idea of a "unitized" mining truck. One of those companies was WABCO. The Model 170 Coalpak utilized a frame design that was an integral part of the entire truck's structure. This allowed the Coalpak design to be more compact, but still retain a rather significant payload capacity, in this case 170 tons. The Coalpak featured the Hydrair II oleo-pneumatic suspension system, with front wheels capable of almost 90 degree steering, giving the coal hauler superb maneuverability in relatively narrow mining pits. Overall length of the unit was 56 feet, 6 inches, with a width of 22 feet, 9 inches. WABCO first released detailed information on its 170 Coalpak in April 1977, with the first unit completed in February 1979 (pictured).

The WABCO Model 170 Coalpak was designed around a diesel-electric drivetrain, with the main power module mounted at the rear of the unit. Standard engine for the Coalpak was a Detroit Diesel 16V-149TI engine, with the availability of an optional Cummins KTA-3067-C unit. Powerplants were 16-cylinder engines rated at 1,600 gross/1,450 net horsepower each. The General Electric drive system utilized a GE GTA-15 alternator and proven 776 electric traction motors in the rear wheels. Tire size for all wheels was 36.00-51, 50PR series type. Empty weight of the Coalpak was approximately 233,000 pounds. The Coalpak was engineered in such a way that 80% of its components were interchangeable with the company's rear-dump 170C Haulpak design. Image date is February 1979.

The first (and only) fleet of 170 Coalpak trucks would ship to Kerr-McGee's Jacobs Ranch Mine in Wright, Wyoming, with the first trucks delivered up and running at the mine by October 1979. They would remain in active service at the mine until they were de-commissioned in mid-1996, with most sold off by the end of that year. Reasons for the failure of the Coalpak, along with other unitized bottom-dump trucks designs of the time (not including Kress) were many. But the worldwide economic recession in the early 1980s, and the advent of larger two-axle, rear-dump models fitted with huge coal bodies were right at the top of the list for most likely reasons. Shown at the Jacobs Ranch Mine in May 1996 is a 170 Coalpak. Note the extra air intake box structures on the upper sides running the length of the unit (both sides), which were added at the mine to increase airflow to the engine module at the rear. *Image by the author*

Chapter 2
Dresser and Komatsu

During the troubled economic periods of the 1970s and 1980s, many of the heavy-equipment manufacturers were stressed in ways never before seen in North America. In the 1970s it was the oil embargos, and then in the early 1980s it was the worldwide economic recession. Companies that were not on firm financial grounds, or were not suitably diversified into other business interests, soon found themselves as possible acquisitions by other large conglomerates. During this time period Dresser Industries of Dallas, Texas, saw this as a perfect opportunity to diversify its company holdings and enter the heavy equipment manufacturing sector as it related to the earth-moving marketplace. Though primarily known as a well-respected manufacturer and supplier of oil-field supplies and services, it was not well known outside of the oil and natural gas industries. But that was soon about to change in a big way.

In 1880, an oil business man by the name of Solomon R. Dresser, in Bradford, Pennsylvania, was granted a patent on May 11 of that year for the design of a unique clamping mechanism that effectively held the rubber packer seals on to their collars in oil wells. Simply known as the "Dresser Cap Packer," it was marketed to oil fields primarily in Western Pennsylvania. This relatively simple product produced by S. R. Dresser & Company was the cornerstone invention of what would eventually become Dresser Industries. In 1885 Dresser received another patent on a "self-packing" coupling for use on natural gas pipelines. This creation of a flexible, leak-proof coupling made possible the future transmission of natural gas through long distance pipelines, but it would take several years for the natural gas industry to adopt Dresser's new coupling in earnest. Finally, in 1898 with the introduction of an insulated coupling, Dresser's company took off with customers in the natural gas industry.

Dresser continued to thrive over the coming years and in 1928 evolved from a privately owned company to that of a public one, operating under the name of the S. R. Dresser Manufacturing Company. In late 1938, the company merged with Clark Bros., a well-respected manufacturer of industrial compressors, forming the new corporation simply known as the Dresser Manufacturing Company. To better reflect the company's standing in the business world and all of the firms and subsidiaries that now made up the parent company, the name of the corporate entity was officially changed in June 1944 from the Dresser Manufacturing Co. to that of Dresser Industries, Inc.

To help cope with the various boom and bust market cycles in the oil and natural gas industries, Dresser made a conscious corporate effort to increase its diversification into other related marketplaces, such as mining and industrial machines, in the 1960s. One such effort in 1967 was the company's attempted buyout of Link-Belt, a well-respected name in the production of excavators and mobile cranes in the heavy-equipment marketplace (through a stock tender offer to the company's shareholders). Link-Belt management did not like the Dresser deal from the beginning and eventually FMC would come out the victor in a bidding war for the company. Undeterred, Dresser Industries continued to look for investment opportunities that they felt could fit within the family of companies owned by them.

In 1972, Dresser closed on a deal with WABCO that saw it taking ownership of the Pneumatic Equipment Division (former LeRoi), which gave the company an outstanding line of compressed air tools and portable compressors. To help in the transition of the purchase, Dresser reinstated the use of the well-known LeRoi

In January 1987, Dresser officially introduced two new "small" Haulpak models, the 140M and 210M. Both of these quarry-sized trucks were of a mechanical-drive nature, and were designed and built at their Peoria, Illinois, operations. Pictured in September 1987 is the smaller of the two models, the 140M.

name (including its old lion-logo) on the product lines.

Dresser made another big move acquisition-wise in May 1974 when it acquired the Jeffery Galion Company of Columbus, Ohio. Jeffery Galion produced Jeffery underground coal mining equipment and Galion-branded motor graders and compactors primarily for the road machinery market. Jeffery Galion's company origin is actually the story of two separate firms—Jeffery Manufacturing Company and Galion Iron Works. Jeffery was originally established in 1887 by Joseph Andrew Jeffrey out of what was left of the old Lechner Mining Machine Co. (established in 1876). Galion came into being in 1907 in Galion, Ohio. In 1928 Jeffery purchased the assets of Galion and formed the Jeffrey Galion Company. Up until the time of the Dresser purchase, Jeffrey Galion had been owned by the Jeffrey family.

Another Dresser purchase concerning mining and construction equipment would involve the Marion Power Shovel Company. The Marion Power Shovel Company of Marion, Ohio, was originally founded in 1884 by Henry M. Barnhart. Over its history, Marion produced some of the world's largest excavating machines, such as walking draglines and stripping shovels for the coal mining industry, large mining cable loading and hydraulic shovels, and blast hole drills. Dresser saw this as a great addition to its mining equipment concerns and finalized the purchase of the company in 1977.

What really placed Dresser into the big-leagues of construction equipment manufacturing was the acquisition of the Construction Equipment Division of International Harvester (IH) in 1982. International Harvester had been having financial problems with its construction division for some time. During the 1970s IH tried in vain to interest other companies into taking the construction group off their hands, but to no avail. Finally in the early 1980s, forced by the world-wide economic recession, IH would be backed into a corner over its ailing division financially. During this time period, Dresser actually tried on a few occasions to purchase the construction division from IH. However, IH simply wanted too much for the division. Dresser's strategy was to simply wait out the economic troubles of the company until they were forced to give it up, which is what ultimately happened in November 1982. The sale of the Construction Equipment Division of IH (referred to as the PAY line Group from 1976 forward) featured the well-established TD-series of track-type tractors. Also included in the deal was the Hough Division, which included the famous product lines of rubber-tired wheel loaders (PAY loader) and dozers (PAY

The Dresser 140M Haulpak was powered by a six-cylinder, Cummins KTA-19C diesel engine, rated at 485 gross/451 net horsepower. Payload capacity range was listed at 35 to 38.5 tons. Standard tire size was 18.00-33, 28PR (E-3). Total empty weight of the small quarry hauler was approximately 66,300 pounds. The 140M was withdrawn from the product line in the mid-1990s and eventually replaced with the Japanese-built 40-ton capacity Komatsu HD325-6 in mid-1998. Image taken in July 1991.

dozer). Other "PAY line" products included in the deal were the scrapers (PAY scraper), of which only a single elevating-type was eventually marketed. What was not part of the Dresser deal were the four-wheel drive off-highway "Pay hauler" lines. These product lines would now fall under a newly formed group referred to briefly as the International-Hough Division of Dresser Industries, Inc.

The economics of the time that crippled International Harvester was also taking its toll on American Standard's WABCO Construction and Mining Equipment division. By 1983 most of its iconic earthmoving product lines had ceased production, with only the Haulpak mining trucks left to continue on in the marketplace. American Standard wanted out of the mining equipment business desperately. At the same time Dresser Industries was looking to further its holdings in that sector of business. On June 1, 1984 Dresser officially became the new owner of WABCO Construction and Mining Equipment and its family of Haulpak truck designs.

The larger of the two mechanical-drive Haulpak quarry trucks built by Dresser was the 210M. The 210M looked much like its little brother, the 140M, but was proportioned slightly larger to handle a 55- to 60-ton payload rating. Pictured in December 1986 is the prototype 210M at the Peoria manufacturing plant.

The Dresser 210M Haulpak truck was powered by a six-cylinder, Cummins KTTA-19C diesel engine rated at 675 gross/641 net horsepower. Standard tires were listed as 24.00-35, 36PR (E-3) type. Empty truck weight was approximately 87,100 pounds. The 210M sold far better in the marketplace than the 140M, but would eventually be removed from production in favor of the Japanese-built 61-ton capacity Komatsu HD465-5 in 1998. Shown is a 210M at work in July 1988.

With the acquisition of the Construction and Mining Equipment division of WABCO, Dresser now had a full line of heavy-equipment offerings under its ownership. The WABCO Haulpak truck line, along with the International-Hough, Marion, and Jeffery Galion product lines, all were now grouped under the Mining and Construction Equipment Division (Group) of Dresser Industries, Inc. Initially, Dresser continued on with many of the old product names still being represented on the machines, such as WABCO. However, starting in early-1986 Dresser made an effort to bring all of the product lines under one common brand name (Dresser) and color scheme (dark yellow with blue strips). Particular brand identities, such as Haulpak and PAY loader, were retained by the company because of their instant recognition in the marketplace. But all did not go as Dresser would have hoped sales-wise. The heavy-equipment industries were slow to rebound coming out of the recessionary times of the early 1980s. Also, the dealer network selling these diverse product lines under the Dresser name were just adequate in North America, but very weak internationally. What Dresser needed was an ally in this ever-expanding global marketplace. And in 1988 that ally (or savior) would be Komatsu.

Komatsu Ltd. can trace its ancestral roots back to the Takeuchi Mining Company of Japan from 1894. In 1917 Takeuchi created the Komatsu ironworks, builders of mining-related equipment and machine tools. In May 1921 Komatsu ironworks was separated from Takeuchi Mining, forming a new company, Komatsu Ltd. As the decades progressed, Komatsu Ltd. continued to grow in Japan and other Asian marketplaces. The company's first true business ventures with North American firms began in 1961. First, Komatsu formed a joint venture with Cummins in the production of engines in Asia. In that same year Komatsu made another agreement with Bucyrus Erie in the production of a limited range of BE excavators to be built in Osaka, Japan, under the corporate name of Komatsu-Bucyrus, K.K. (production starting in 1963 and ending in 1981). In late 1966 Komatsu had agreed to supply track-type tractors to WABCO in North America starting in January 1967. WABCO desperately needed a line of tracked dozers in its product line, and Komatsu saw this as a perfect opportunity to break into the North American marketplace. But dealer and customer resistance to these Japanese-sourced machines caused the agreement between the two companies to be abandoned by the end of 1969. Lessons learned from these joint ventures and marketing agreements led Komatsu Ltd. to form a North American subsidiary headquartered in the United States—Komatsu America Corporation.

Early on it was Komatsu America Corporation's priority to establish a reliable dealer network in the United States. As the 1970s progressed, the monetary exchange rates for the U.S. dollar and Japanese currency tipped in favor of the yen, making the Komatsu offerings far more tempting to prospective customers. And, just as

Not long after Dresser Industries acquired the Construction and Mining Equipment Division of WABCO, they started to rebrand the models in 1986. One of those was the WABCO 85D, which was now referred to as the Dresser 325M Haulpak. Other than the nomenclature, the 325M was the same tried and true 85- to 95-ton capacity mechanical-drive hauler the mining industry had come accustom to over the years. Standard power for the truck was supplied by a 12-cylinder Cummins KT38-C diesel, rated at 925 gross/876 net horsepower. An 8-cylinder Detroit Diesel 8V-149TI with an output of 900 gross/858 net horsepower was also available. Empty weight was listed at approximately 127,000 pounds. Image taken in December 1988.

the strong U.S. dollar helped Komatsu and its dealers in the 1970s, a weakening of it and the increasing value of the yen in the mid-1980s would hurt it. To help in the fluctuations of the exchange rates between the two marketplaces, Komatsu needed to establish manufacturing capabilities in North America to off-set its export business from Japan. In 1985 the company purchased a 55-acre empty plant in Chattanooga, Tennessee, with production officially getting underway in 1986. The new production plant, operating as Komatsu America Manufacturing Corp., produced approximately 20 models of construction equipment, including medium-sized hydraulic excavators and large-sized wheel loaders. However, even more manufacturing space would be needed by Komatsu, and the opportunity to make its business expansion feasible would come about in 1988 with Dresser Industries.

In early 1988 Dresser Industries and Komatsu Ltd. signed a "memorandum of agreement" to create a joint-venture company to engineer, manufacture, and market construction and mining equipment in the Western Hemisphere. On September 1, 1988 the deal officially went into effect, creating a new wholly-owned subsidiary company of both Dresser Industries and Komatsu Ltd., known as the Komatsu Dresser Company (KDC). The new 50/50 joint venture company would be made

Komatsu Dresser Co. originally introduced their mechanical-drive 93- to 100-ton capacity range hauler, the Dresser 330M Haulpak, in 1991. Based largely on the Komatsu HD785-3 series produced overseas, the Dresser version was assembled in Peoria and sold in the Americas as the 330M. Standard engine choice was the 12-cylinder Cummins KTA38-C, rated at 1,050 gross/1,001 net horsepower. Empty weight of the truck was approximately 149,716 pounds. By 1995 the Dresser name was replaced with Komatsu solely. Image date is October 1995.

The Komatsu 330M Haulpak was well received in the marketplace and proved to be a solid performer in the field. The 330M chassis also made a good tractor variation for coal hauling duties. Shown working at the BHP LaPlata Mine, located just north of Farmington, New Mexico, in November 1996, is a 330M equipped with a set of MEGA Magnum Tandem CH120 coal bottom-dump trailers, rated at 240 tons combined. *Image by the author*

up of all of Dresser's construction and mining equipment product lines, with the exception of the Marion and Jeffery Divisions, and the air tools business, which would remain part of Dresser Industries Mining and Construction Equipment Division. Komatsu contribution to the deal would be its Chattanooga assembly plant and Brazilian operations. Both entities would maintain their existing dealer networks and continue to sell their product lines separately under the Dresser and Komatsu trademarks and colors.

Though the 50/50 joint venture company must have made sense to both Dresser and Komatsu at the time, to the marketplace and potential customers, KDC's product lines seemed confusing and at odds with each other. Both companies offered competing wheel loader, excavator, and track-type tractor models through two dealer networks, quite often to the same customer. The one exception was the Haulpak line of mining trucks. Komatsu did produce a line of off-highway haulers, the largest of which were the 132-ton capacity HD1200-1/HD1200M-1 and the 176-ton HD1600-1, but the Haulpak models were in a whole different league. To many, the Haulpak line of trucks were some of the best the industry had to offer, and the true star product line of Dresser's side of the joint venture.

The late 1980s and early 1990s were tough financially for the Komatsu Dresser Company. KDC lost money in 1989, 1990, and 1991 (a recessionary year for most). These losses were simply too much for Dresser Industries (and its shareholders) to take. For decades Dresser was, for the most part, a very profitable company. But its strategy to diversify into a capital intensive and very competitive business like construction and mining equipment was not paying off. Finally, in 1992 Dresser had had enough of the losses. On August 1st of that year, Dresser Industries officially cut its ties to the construction and mining segments of its business by spinning off its Industrial Products and Equipment divisions (consisting of the 50% share in KDC, the Marion, Jeffry, and air tool divisions) to its shareholders, forming a separate standalone corporation named Indresco, Inc. With Dresser Industries clear, it was up to Indresco to continue on under mounting losses from its mining and construction divisions. It soon became obvious that Indresco was not up to the task. To raise capital, Indresco sold 31% of KDC to Komatsu Ltd. on September 30, 1993 (announced July 20, 1993), raising its total ownership of KDC to 81%, with an option to purchase the remaining part of the joint-venture company. One year later on September 30, 1994 (announced September 1, 1994), Komatsu exercised its option to the rest of KDC and purchased the remaining 19% of KDC from Indresco, making it the sole owner of the troubled venture. In January 1996 the name of the wholly owned

Komatsu officially changed the nomenclature of the 330M built in North America to that of the HD785-5 (original Komatsu model name of model built in Japan) in the year 2000. In late 2006 the hauler was upgraded to an HD785-7 version. The current mechanical-drive model is powered by a 12-cylinder Komatsu SAA12V140E-3 diesel engine, rated at 1,200 gross/1,178 net horsepower, with a payload capacity rated at 100.3 tons, an empty weight of approximately 158,800 pounds, and a standard tire size of 27.00, R49 series type. The HD785-7 was initially only built in Japan, but starting in March 2007, production began on the model at the Peoria, Illinois, assembly plant as well. Pictured at work in April 2008 is the HD785-7. *Komatsu America Corp.*

subsidiary was officially changed from Komatsu Dresser Company to that of the Komatsu America International Company. In April 2002 Komatsu pushed through a series of restructuring efforts that combined the U.S. construction, mining, and utility operations under the new subsidiary name Komatsu America Corp. This included Komatsu America International, Komatsu Mining Systems (KMS), and Komatsu Utility.

During the late 1990s and early 2000s, Komatsu was steadily refining its holdings into a very cohesive and effective global player in the construction and mining marketplace. The previous KDC joint venture still left a few loose ends for Komatsu that would eventually be worked out. In 1995 Komatsu sold a portion of KDC to Huta Stalowa Wola S.A. (HSW), forming the new joint venture, Dressta Co. Ltd., for the manufacturing of the old Dresser designs in Poland, to be marketed under the Dresser name worldwide. On October 1, 1999, Komatsu discontinued the use of the "Dresser" name on all of the Dresser product lines, due to the termination of the trademark license. All machines from this point forward would carry the Dressta trademark in all marketplaces. In June 2005 the circle was completed when HSW purchased the remaining assets of Dressta from Komatsu, finally freeing it from the past legacy of the Dresser ordeal.

As for Indresco Inc., the company no longer had any ties to Komatsu after the sale of its last portions of KDC in 1994. But life for Indresco improved little even after the sale of the KDC assets. On October 27, 1995, Indresco announced the sale of its Jeffrey Division to a Cleveland, Ohio-based investment group, forming the company Jeffrey Mining Products. On November 1, 1995, Indresco also announced the adoption of a holding company structure with a new corporate name of Global Industrial Technologies, Inc. Marion Division (as it was known under Indresco) would re-acquire its old name, The Marion Power Shovel Company, and become a subsidiary of Global (Marion would eventually be purchased by its longtime rival Bucyrus Erie in August 1997).

Throughout all of these turbulent times, mergers and acquisitions, the Haulpak truck lines and their present-day decedents have survived. The off-highway haulers have been marketed under various corporate structures over the years. These included Haulpak Division (1988) during the Dresser years, and Komatsu Mining Systems, Inc. (1997) during the Komatsu America International Co. time period. Since 2002, the mining hauler program resides under the Komatsu America Corp. banner. Even though the "Haulpak" trade name was phased out of use by the end of 1999 by Komatsu, its legacy and heritage live on in the mighty quarry and mining haulers still being manufactured at the company's Peoria, Illinois operations.

Introduced in late 1986, the Dresser 445E Haulpak diesel-electric drive truck was essentially the older WABCO 120D with a new product line nomenclature to better represent the hauler's loaded gross vehicle weight (in this case 445,000 pounds). The 445E had a choice of two standard engines—the Detroit Diesel 12V-149TIB or the Cummins KTA-38C, both rated at 1,200 gross/1,129 net horsepower. Payload capacity range was listed at 115 to 125 tons, with an empty weight of approximately 195,900 pounds. Standard tire size was the 30.00-51, 46PR type. Image taken in March 1991.

The year 1986 would see the older mechanical-drive WABCO 140DM be re-introduced in the form of the Dresser 510M Haulpak, and also in a new diesel-electric drive version referred to as the 510E. The 510M would quickly disappear from the product line leaving only the 510E Haulpak. Two engines were offered in the form of the Detroit Diesel 12V-149TIB or the Cummins KTA-38C, both rated at 1,350 gross/1,279 net horsepower. Image date is April 1988.

The Dresser 510E Haulpak electric-drive truck was rated with a payload capacity range of 133 to 150 tons, with a gross vehicle weight of approximately 510,000 pounds (empty weight 209,820 pounds). The 510E would stay in the product line as a Komatsu model well into 1997, even after its replacement had been released in 1996 in the form of the 530M, because of overseas demand in certain marketplaces such as Russia. Shown in June 1989 is a 510E equipped with a high-volume coal body.

Komatsu officially unveiled its large mechanical-drive 530M Haulpak truck at MINExpo in September 1996 (in development since 1994), as the eventual replacement for the aging 510E. Rated as a 150- to 165-ton capacity hauler, the first 530M rolled out of the Peoria plant in February 1996. In 2000 the nomenclature was changed on the 530M to that of the HD1500-5 to simplify Komatsu's global product line offerings in all sales regions. Image taken in 1997.

The Komatsu HD1500-5 was replaced by an upgraded HD1500-7 model variation in the summer of 2007. Power for the big nominal 159-ton capacity, mechanical drive truck is supplied by a 12-cylinder Komatsu SDA 12VV160 diesel engine, rated at 1,487 gross/1,406 net horsepower. Standard tire choice is a 33.00 R51 series type rubber. Approximate empty weight is listed at 232,144 pounds. Photo taken in April 2008. *Komatsu America Corp.*

Dresser officially introduced its 630E Haulpak diesel-electric drive truck at the American Mining Congress show in October 1986. The 630E was essentially a combination of the WABCO 170D/190 mining truck platforms. Two 16-cylinder engine choices were available—a Cummins KTTA 50-C or a Detroit Diesel 16V-149TIB. Both powerplants were rated at 1,800 gross/1,704 net horsepower. The 630E utilized a 17-foot, 10-inch wheelbase, with an overall length of 39 feet, 2 inches, and a maximum width of 23 feet. Image date is November 1987.

The Dresser 630E was rated as a 170- to 190-ton capacity haul truck, depending on the type of electric wheel motors ordered (GE 776 or 788 model types). This payload was carried on standard 36.00x51-sized tires. As the 630E mining truck matured in the Haulpak product line, it would receive numerous upgrades, such as revised front wheel castings, new air cleaners, and a redesigned front grille. Average empty weight (with 776 wheel motors) was 250,350 pounds. Maximum loaded gross vehicle weight with 776 wheel motors was 610,350 pounds (630,000 pounds when equipped with 788 motors). In 1995 the 630E became a Komatsu product after the Dresser name was finally dropped from use. The 630E was offered in the Komatsu product line as late as 2001. Shown in July 1991 is a 630E Haulpak being loaded at a West Virginia coal mining operation.

The next rung up in the Haulpak electric-drive mining truck line from the 630E was the popular 685E. First announced by Dresser in 1988, the 685E utilized the same wheelbase as the 630E, and was designed to handle payloads in the 190- to 200-ton capacity range. Empty truck weight was approximately 283,000 pounds, with a maximum loaded weight of 685,000 pounds. This load was carried on 37.00 R57-sized tires. Image taken in August 1990.

The Dresser 685E Haulpak was offered with two 16-cylinder engine choices. Customers could specify the Detroit Diesel 16V-149TIB or the Cummins KTTA 50-C powerplants, both rated at 2,000 gross/1,854 net horsepower. Wheel motor of choice was the tried and true GE 788. By 1997 (now a Komatsu model) only a new Cummins K1800E diesel engine (1,800 gross/1,704 net horsepower) was offered. Length of the truck was 40 feet, with an overall width of 23 feet, 8 inches. Pictured in May 1989 is a 685E at the Barrick Goldstrike Mine, located near Carlin, Nevada.

The 685E Haulpak sold very well for both Dresser and Komatsu, and was considered by many in the mining industry as the premiere diesel-electric drive mining truck in the 200-ton payload class range. Komatsu would introduce the 730E Haulpak in 1995 as the eventual replacement for the 685E, which would remain viable in the product line until 1997. Image taken in May 1989.

Komatsu officially announced its 730E Haulpak electric-drive mining truck in 1995. Initially rated with a maximum payload capacity of 212 tons, it would be reduced to a range of 195 to 205 tons the following year. Though the 730E was built as a replacement for the aging 685E, both were available at the same time in the product line for a few years. The 730E was designed around a 19-foot, 4-inch wheelbase. The first prototype 730E was completed in December 1994. Image date is 1997.

The Komatsu 730E was originally powered by a Cummins K2000E diesel engine, rated at 2,000 gross/1,860 net horsepower. As of 2011, the standard engine of choice is the Komatsu SSA 16V159. Power ratings are the same as the earlier K2000E offering. Standard wheel motors are the highly reliable GE 788 type. Empty weight of the 730E is approximately 309,950 pounds, with a nominal loaded gross vehicle weight of 715,000 pounds. Standard tire size is 37.00 R57 series type. Overall length is listed at 42 feet, 1 inch, with a width of 24 feet, 9 inches. An optional 730E Trolley edition of the truck is also offered by Komatsu. Image taken in April 1997.

The 830E Haulpak is truly one of the greatest diesel-electric drive haul trucks ever produced by any manufacturer. For years it dominated the 240-ton payload class of mining trucks in the marketplace, and is the all-time best selling 240-ton class, diesel-electric drive model of all time. Originally referred to by Dresser as the 780E while the truck was under development, the 830E Haulpak was first announced by the company in November 1987. It would make its official world debut at the American Ming Congress show, held in Chicago, Illinois, in April 1988 (pictured).

Everything about the Dresser 830E Haulpak was "big" for its day. From payload to power, the 830E had it all. The 830E was designed around an extremely strong frame, utilizing a 20-foot, 10-inch wheelbase. Engine of choice was the 16-cylinder Detroit Diesel 16V-149TIB, rated at 2,200 gross/2,054 net horsepower. Electric wheel motors specified for such a large truck were a pair of GE 787 units. Empty weight of the hauler was approximately 318,569 pounds, with a maximum loaded gross weight of 830,000 pounds. Payload capacity was 240 tons, which was considered huge for a two-axle truck. Tires were big 40.00-57, 68PR series type, which were the largest available for a haul truck at the time. Image taken in July 1988.

In late 1990 Dresser introduced the "2nd Generation" 830E Haulpak truck with a host of improvements to greatly increase its overall productivity. The quickest way to recognize one of these upgraded models was the elimination of the front-mounted air-cleaner covers. The company also started field trials in late 1991 on a more powerful engine option for the 830E at an iron ore mine in Michigan, with trucks equipped with Detroit Diesel's new 20-cylinder 20V-149TIB engine. This big diesel was capable of delivering 2,500 gross/2,334 net horsepower when installed in the 830E. Further field-follow trucks would be shipped to copper mining operations in Arizona over the next two years. These early trucks featured a longer nose section to accommodate the larger engine. Pictured in September 1996 is one of these more powerful 830Es operating at the Asarco Mission Mine. *Image by the author*

Today's Komatsu 830E is as popular as ever in the mining industry. The 830E is offered with a long options list, from railings and steps, to custom specified dump box selections. Whatever the customer wants, Komatsu can supply the 830E with it. In the late 1990s the MTU/DDC 16V4000 (rated at 2,500 gross/2,409 net horsepower) engine was installed in the 830E. Today, the Komatsu SDA16V160 (rated at 2,500 gross/2,360 net horsepower) is the engine of choice. Current payload capacity range is 240 to 255 tons. Empty weight is approximately 358,259 pounds, with a loaded nominal weight of 850,650 pounds. Overall length is listed at 46 feet, 5 inches, with a width of 24 feet. Shown in July 2003 is an 830E equipped with a customer-specified Trinity Industries, Inc. T-Max lightweight dump body. *Image by the author*

At the September 2004 MINExpo show in Las Vegas, Nevada, Komatsu officially unveiled its 830E-AC electric-drive truck. In this variation of the company's popular 830E, the model features a GE AC electric drive system (as opposed to the standard model's DC drive system) utilizing GEB25 motorized wheels. Engine is the same as that in the standard model, the Komatsu SDA16V160 (rated at 2,500 gross/2,360 net horsepower). Standard tire size is a 40.00R57 series, but optional larger 46/90R57 units are available. Empty weight is listed at 362,000 pounds, with a loaded nominal weight of 850,650 pounds. Payload range is 244 to 250 tons in capacity. As of March 2011, Komatsu (and Dresser) had shipped an impressive 1,466 830E trucks worldwide (both DC and AC). Of this total, almost 600 of these are the newer AC-drive version. *Image by the author*

Komatsu's all new 860E-1K electric-drive truck was officially made public at the 2008 MINExpo, held in September in Las Vegas. The nominal 280-ton capacity hauler features the company's new state-of-the-art Komatsu AC Electric Drive System (also referred to as Komatsu-Drive) designed by the company, powered by the latest Siemens control package. The unique, liquid-cooled IGBT AC-drive control system from Siemens provides the 860E-1K with a smooth delivery of torque and traction. The new model is powered by a Komatsu SSDA16V160 Tier 2 diesel engine, rated at 2,700 gross/2,550 net horsepower. Tire choices are standard 50/80 R57 series, with optional 50/90 R57 available. Empty weight is listed at 441,700 pounds, with a maximum loaded gross vehicle weight 1,001,700 pounds (nominal loaded 987,700 pounds). The hauler utilizes a 20-foot, 8-inch wheelbase, with an overall length of 49 feet, and a width of 30 feet, 10 inches. Full scale production of the 860E-1K commenced in late 2009. *Komatsu America Corp.*

Komatsu Dresser Co.'s (KDC) first AC-electric drive truck release was the ground-breaking 930E Haulpak. Announced in 1995, the giant Komatsu introduced many industry firsts to a mining truck design, such as an efficient GE-designed AC electric drive system. Another first for the model included the use of massive 48/95 R57-sized tires, the largest ever made for an off-highway hauler at the time. The huge 285- to 310-ton payload capacity was also of significant merit for a two-axle truck design. Image taken in August 1995.

The original series of Komatsu 930E Haulpak AC-drive trucks were powered by a 16-cylinder MTU/DDC 16V396 TB44L diesel engine, rated at 2,682 gross/2,500 net horsepower. Approximate empty weight of the hauler was listed at 411,300 pounds, with a maximum loaded weight of 1,034,000 pounds. The 930E was designed around a chassis featuring a 20-foot, 10-inch wheelbase. Pictured in October 1997 is the first field-follow 930E Haulpak truck built by KDC at Fording River Coal, located in British Columbia, Canada. Start-up of this truck was in 1996. *Image by the author*

In 1997, Komatsu made some very significant changes to the 930E, which would raise its maximum payload capacity to 320 tons. To make this possible, a new MTU/DDC 16V4000 series diesel engine was installed, which was rated at 2,700 gross/2,500 net horsepower. Larger 50/90 R57 radials were also fitted to handle the extra capacity. All of these changes did add a bit more weight to the truck as a whole. Empty weight was now listed as 419,470 pounds, with a maximum loaded weight of 1,059,000 pounds. Image taken in March 1997.

In the summer of 1999, Komatsu announced that it was delivering an improved version of its 930E equipped with larger 53/80 R63 radial tires mounted on 63-inch diameter rims to the Barrack Goldstrike mine in Nevada. Often referred to as the "Bigfoot" truck, it would be reclassified as the 930E-2 in November 1999. Though capacity remained unchanged at 320 tons, maximum gross vehicle weight increased to 1,100,000 pounds. Shown is one of the 930E-2 Barrick trucks equipped with a Trinity Industries T-Max lightweight dump body in November 2004. *Image by the author*

Along with the larger tire and wheels fitted to the Komatsu 930E-2, a new engine option was also added. In addition to the MTU/DDC 16V4000, a 16-cylinder Cummins QSK60 diesel engine was now available with power ratings of 2,700 gross/2,550 net horsepower. Empty weight of the truck equipped with the MTU was listed at 442,214 pounds, and 446,034 pounds when fitted with the Cummins powerplant. Photo taken in June 2000. *Image by the author*

The Komatsu 930E series has proven itself time and time again in some of the largest mining operations around the word. During the years Komatsu has continually improved and updated the model line. In 2004 the 930E-2 became the 930E-3, which brought with it a new standard Komatsu SSDA16V160, 16-cylinder diesel engine (based on the Cummins QSK60), rated at 2,700 gross/2,550 net horsepower. In 2007 this model was upgraded to the 930E-4, which utilized the same engine, but was now Tier 2 emissions compliant. Maximum payload capacity of all these models has remained pegged at 320 tons. Overall length is 51 feet, 2 inches, with a maximum width of 29 feet, 10 inches. Shown in June 2000 is a 930E-2 being loaded at a coal mine located in the Powder River Basin of Wyoming. *Image by the author*

Komatsu first introduced its high-horsepower version of the 930E-2 at the October 2000 MINExpo in Las Vegas, Nevada, after months of testing at its proving grounds at the old Twin Buttes mine site located south of Tuscan, Arizona. Originally referred to as the 930E-2SE, the AC electric drive truck utilized a powerful 18-cylinder engine that gave the hauler the extra power needed for extreme mining applications where haul road truck speeds needed to be maintained in high altitude and (or) deep-haul pit operations, such as those commonly found in copper mines in South America. Other upgrades to the model line included the 930E-3SE in 2004 and the 930E-4SE in 2007. All versions utilized the large Komatsu SSDA18V170 diesel engine (based on the Cummins QSK78), rated at 3,500 gross/3,429 net horsepower. Payload and tire sizes were the same as the standard 930E-4 model. Empty weight was listed at 474,670 pounds, with a loaded maximum gross vehicle weight of 1,115,000 pounds. Pictured in June 2008 is a 930E-4SE equipped with a customer specified DT Hi-Load dump body. *Komatsu America Corp.*

In 2005 Komatsu started field-testing of its latest ultra-hauler, the 960E-1. The new 360-ton capacity electric-drive mining truck grew out of the engineering experience gained from the 930E-3SE program, such as the use of the Komatsu/Cummins jointly-designed 18-cylinder Komatsu SSDA18V170 engine package (3,500 gross/3,346 net horsepower). The 960E-1 utilizes a GE sourced AC-electric drive system incorporating GDY 108 wheel motors. Tires on the big hauler are huge, low profile 56/80 R63-sized radials. Empty weight of the 960E-1 is approximately 550,000 pounds, with a nominal loaded gross vehicle weight of 1,270,000 pounds. Load capacity is a nominal 360 tons. Wheelbase is listed at 21 feet, 10 inches. In October 2011 Komatsu introduced an upgraded 960E-2 model. Shown at the Jacobs Ranch Mine, located south of Gillette, Wyoming, on June 14, 2005, is the first 960E-1 field-follow unit during assembly. This truck would carry its first load at the mine on June 16. *Image by the author*

In 2006 Komatsu would start the field-testing program of its new 960E-1K electric mining truck, utilizing the company's new Komatsu Drive System incorporating Siemens power components. The new AC electric drive system was designed by Komatsu in cooperation with Siemens Group Industrial Solutions. The AC electric drive systems are the major design difference between the 960E-1K and the 960E-1, which utilizes GE drive technology. Payload, engine, power output, and weights are the same as those of the 960E-1. Length of the model is 51 feet, 2 inches, with a maximum width of 31 feet, 6 inches. Shown in July 2006 is the pilot 960E-1K equipped with Komatsu-Drive in Fort McMurray, Alberta, just before delivery to its new home at Suncor in the oil sands, to start its long-term field-follow trials. The 960E-1K was officially released for worldwide sales in the summer of 2010. In October 2011 an upgraded 960E-2K was introduced. *Komatsu America Corp.*

In 1958, preliminary meetings were first held between General Electric, Hanna Mining Company, and Unit Rig & Equipment Co. on the feasibility of producing a diesel-electric drive mining hauler. By mid-1959 an agreement was reached on a truck design built to the specifications requested by Hanna Mining. In January 1960 Unit Rig's revolutionary electric-drive ore hauler, the Lectra Haul M-64, was officially unveiled to the industry. Though only a prototype design, the M-64's use of GE's motorized wheel drive system would point the way not only for future Lectra Haul designs, but the mining industry as well. Pictured is the pilot M-64 (Model No. 21A-64 PRD Ore Hauler, S.N. 51) in late January 1960.

Chapter 3
Unit Rig

The origins of Unit Rig are similar to that of Dresser, in the fact that both were born in the oil field equipment industries, and both would enter the off-highway truck marketplace in a need to diversify. But the approach both companies took to accomplish this diversification was quite the opposite. In the case of Dresser, they chose the acquisition of another company (WABCO). In the case of Unit Rig, they relied on the talents of their seasoned in-house engineering department.

Unit Rig can trace its origins back to 1935, when an oil-field salesman by the name of Hugh S. Chancey conceived an idea for a rotary drilling rig that was easier to set up and move to different working locations. In the design the draw-works and engine module would be mounted on a single common frame, making it a unitized drilling rig. In early 1936, Chancey, along with two other business men, Jerry R. Underwood and William C. Guier, entered into a partnership for the creation of two companies, Unit Rig & Equipment Company, and Portable Drilling Co., Inc. In the creation of two companies, Unit Rig would be responsible for the design and building of the unitized rig, with Portable Drilling actually placing it into active service. Unit Rig would establish a location to manufacture the rig, referred to as the U-10, in West Tulsa, Oklahoma. The first U-10 unitized drilling rig was completed by Unit Rig in 1937, and was successfully placed into service by Portable Drilling, drilling an oil well just outside Oklahoma City. The success of the U-10 Portable Rotary Drilling Unit would lead to more equipment designs targeted for the oil fields, such as the U-16 Combination Rotary and Cable Tool Rig, the U-17 Service Hoist, the U-19 Auxiliary Spudder, and the U-20 Telescoping Super Mast. With the success of these new products in the industry,

The Lectra Haul M-64 was powered by a 12-cylinder Cummins VT-12-BI diesel engine, rated at 700 gross/625 net horsepower. The main components in the electric drive system consisted of a GE GT-594 generator and GEZ 5319 traction motors, mounted in all the wheel assemblies, giving the hauler full four-wheel-drive capabilities. The M-64 was designed around an articulated frame design utilizing Goodyear 37.5-33, 48PR series tires, which also doubled as the main suspension system for the truck. Payload capacity was 64 tons in its 40-cubic-yard struck rear-dump body, with an approximate overall empty vehicle weight of 112,000 pounds. Length of the truck was 41 feet, 8 inches. The Lectra Haul M-64 is shown here in late January 1960 being tested at Standard Industries, Inc. rock quarry, located east of Tulsa, Oklahoma.

Unit Rig would move its offices and assembly operations to new locations in downtown Tulsa in 1941.

Business was going well for both companies, but cracks in the partnership between the three men responsible for the creation of Unit Rig and Portable Drilling would come to a head in 1946, when court proceedings were started to incorporate the two firms and eliminate the partnership. In 1947, Jerry Underwood died, leaving Chancey and Guier to fight for control of the two firms in court. In the end, Guier would gain complete control of Unit Rig, with Chancey getting control of Portable Drilling. Because of the bitter court battle between the two surviving founders, it became necessary for the two companies to go their separate ways when Unit Rig refused to work with Portable Drilling.

On January 22, 1951, Unit Rig & Equipment Co. was sold to Kenneth W. Davis, Sr. of Fort Worth, Texas. Davis was no stranger to the oil field business. Companies such as Mid-Continent Supply Company, and Loffland Brothers Drilling Company, were well known firms in the oil field supply business. These companies, along with a host of others owned by Ken W. Davis, Sr., would eventually become part of Kendavis Industries International, Inc.

In August 1956, Jesse L. Vint, Jr. became President of Unit Rig. Over the years Vint had served as Chief Engineer, V.P. of Engineering, and V.P. President of Sales for the company. He would remain the head of Unit Rig until his retirement in April 1982. It was under Jesse Vint's tenure at the helm that Unit Rig would branch out and diversify itself away from oil field equipment production and into the off-highway mining truck marketplace, and grow the company into one of the industry's leading suppliers of diesel-electric drive haul trucks.

In January 1958, another of Ken Davis, Sr. companies, Mid-Continent Supply Co. of Fort Worth, Texas, became the exclusive distributer for Unit Rig's oil field product lines. Other non-oilfield business opportuni-

Not long after the M-64 started testing at the Standard Industries quarry location, problems started to show up in the truck's design, mainly revolving around the failure of the gooseneck hitch that connected the tractor unit to the rear dump portion of the truck. After a redesign of the hitch, the prototype M-64 was shipped to Hanna Mining in Minnesota to continue its long-term field trial evaluation tests. The truck was assembled in Hibbing and then driven 20 miles to its first working location at Hanna's Hunner Mine. It would later be moved to Hanna's Pierce Mine for further evaluations. Shown is the Lectra Haul M-64 around June 1960, as it gets ready to make its way to Hanna Mining.

ties for Unit Rig at the time included the awarding of a contract from the U.S. Government for the production of a "scissors-type" folding bridge and bridge launching system for the U.S. Army Corp of Engineers consisting of an aluminum, folding bridge and launcher (available in 43- and 63-foot sections) mounted to an M48 medium tank chassis (Class 60, AVLB), in 1958. In 1959, the company was awarded another contract for the design and production of a high-speed combat entrenching vehicle for the U.S. Army, with the first completed unit tested in May 1960. These military contracts carried the company during the slow times in the oilfield business. But what the company really needed was the production of some kind of a product that was outside of the oilfield industry and limited government contracts, which could make good use of Unit Rig's manufacturing facilities. Unit Rig looked toward the mining marketplace and the off-highway hauler.

During the late 1950s, mining companies were starting to investigate the feasibility of electric-drive haul trucks by means of overhead trolley lines, or electric-drive motors powered by a separate diesel engine. R.G. LeTourneau, Inc. had combined the electric wheel motor and its planetary drive in the wheel assembly itself back in 1950, making it a very compact and efficient design. In 1959, they would build a prototype 65-ton capacity truck design, the Series TR-60, to the specifications put forward by the Anaconda Company, for use at its Berkeley Pit in Butte, Montana. General Electric was also working on a wheel motor design for the mining industry, borrowing from their vast knowledge from building locomotive drive systems. GE first approached Unit Rig in 1958 on the possibility of producing a diesel-electric drive mining hauler powered by the new GE electric wheel system. Also involved in the negotiations was the Hanna Mining Company. The truck would be built to the design specifications put forth by Hanna, who would be the recipient of the truck. Unit Rig was intrigued by the offer, and after months of industry analysis, agreed to build a prototype hauler for Hanna utilizing the GE drive system in June 1959.

In January 1960, Unit Rig had completed its prototype mining truck, the Lectra Haul M-64 Ore Hauler. The M-64 was configured as a two-axle, all-wheel drive design, utilizing a separate tractor/trailer arrangement. A major design feature of the hauler was the use of Goodyear Tire & Rubber Co. low-pressure tires that also doubled as the suspension system. Goodyear insisted that the design was sound for the intended application. But it soon became evident that a mining hauler of this type needed something far more than the tires to act as the suspension system. Not long after deployment, the gooseneck failed that attached the trailer to the tractor hitch. This required a complete redesign of the gooseneck. After this was accomplished, the prototype M-64 was shipped off to Hanna Mining in June 1960. Over the next year or so, the M-64 was worked at two separate Hanna Mining locations. Still, like before, the gooseneck would show itself as the weak spot in the truck's design. Continued failures of the gooseneck caused the M-64 design to be abandoned by all concerned. The initial failure of the truck was due more to the use of the tires as a suspension system. The pounding and extra stress placed on the frame and hitch was enormous, and the ride and handling were not much better. Unit Rig engineers would go back to the drawing boards to design a new truck layout utilizing the GE system.

Unit Rig engineers were convinced that the diesel-electric drivetrain layout utilizing the GE motorized wheel system was sound. Unit Rig quickly refocused its engineering efforts on a new chassis design. For inspiration, they looked at the current LeTourneau-Westinghouse designs, which were fast becoming the accepted layout of future mining haulers. The end result was a completely modern design utilizing a two-axle design with rear-driven wheels only. The design would also incorporate a rubber cushioned suspension system referred to as "DYNA-FLOAT" by the company. In the summer of 1963 Unit Rig unveiled their new off-highway mining hauler design—the M-85 Ore Hauler. The M-85 was truly a game changer for Unit Rig & Equipment Co., and for the surface mining industry as a whole. Here was a truck equipped with a state-of-the-art drivetrain with a chassis to match. The M-85 would propel Unit Rig into the mining industry and away from the oilfield business that had originally given birth to it. For the company it was the beginning of the era of the "Lectra Haul."

After the successful trial testing of a handful of Lectra Haul M-85 trucks in late 1963 and early 1964, the M-85 would go into full production for Unit Rig in May 1964. After that there was no looking back for the company. Orders started to flow in from some of the largest copper and iron ore mining companies in North America. Companies such as Anaconda, Cyprus Pima Mining, Kennecott Copper, and Molybdenum Corp. were all early adopters of the innovative diesel-electric drive M-85. Unit Rig would follow up the M-85 with the introduction of the M-100 in 1965, and the M-120 in 1968. All of these truck models shared the same basic 15-foot wheelbase frame design (reinforced in key areas for strength on the larger capacity models), which was one of the most trouble-free and reliable chassis layouts ever created by the company.

In the late 1960s, Unit Rig tried its hand at producing other types of heavy industrial equipment designed around a diesel-electric drive platform. In 1968 the company introduced a 60-ton capacity fork-lift, the L-120 Lectra Lift. In 1969, the T-150 Lectra Haul was placed into service as a high capacity tow tractor designed for use with Boeing's 747 passenger jet airliner. Unfortunately, the Lectra Lift program was a complete failure (with only two produced), while the tow tractors (the T-150 and the smaller T-90 first announced in 1970) were only moderately successful for the company (69 were built of all versions).

Unit Rig was the leading supplier of large capacity mining trucks from the mid-1960s and throughout most of the 1970s. During this time period Unit Rig's biggest competitor was WABCO. Companies such as Euclid and Caterpillar simply did not offer anything that was comparable to the Unit Rig lines at the time. Terex offered its 33-15 series diesel-electric drive hauler as a direct competitor to the Mk 36, and even though

The M-64 ore hauler's GE drive system performed well in the field. But the same could not be said for the truck's articulated frame. Again, gooseneck failures plagued the unit throughout its testing program, with two more reported major rebuilds required to keep the prototype operable. These gooseneck failures could be directly attributed to the fact that the M-64 did not have a true suspension system to speak of. Instead, it relied on the low-pressure Goodyear tires to act as a suspension system. In the end the constant daily pounding of the truck in service was too much for its frame design to handle. Shown at work in one of Hanna Mining's properties in late 1960, is the M-64. Note the reinforced gooseneck design now being utilized on the truck at this time in its testing program.

the 33-15 series sold well, the Mk 36 sold in far larger numbers. This is not to say that all of the Unit Rig designs were perfect—they were not. The 17-foot wheelbase frame designs of the M-120.17 and Mk 30, and the 17-foot, 6-inch wheelbase layout of the Mk 33 and Mk 36, were prone to failure in the field. Unit Rig stood behind these trucks and repaired them as needed in the field, but the warranty claims were substantial. Nevertheless, the models still sold well for the company, especially the Mk 36, and were considered exceptional haulers in the marketplace, even with the frame issues.

The recessionary times of the early 1980s were not as difficult for Unit Rig as they were for other heavy equipment manufacturers. Sales were down, but not to a point where it affected the company's operations in a vastly negative way. As things started to look up in the world economies, Unit Rig was in a good position for making possible acquisitions to help round out its product lines and better align its offerings to the mining marketplace. On March 30, 1984, Unit Rig closed the deal on the purchase of Dart from PACCAR, builders of the Peterbilt and Kenworth truck lines. PACCAR wanted to rid itself of the slow selling Dart model lines and concentrate on its over-the-road truck line sales,

With just over 3,000 hours of operation under its belt, the pilot Lectra Haul M-64 was shipped back to Unit Rig's Tulsa plant in 1961 for a complete engineering evaluation. Unit Rig engineers discarded the articulated frame design, but kept the diesel-electric drive powertrain set-up. The new Lectra Haul truck program would consist of a more conventional rigid-frame, two-axle design, with only the rear wheels driven. It would also be able to carry an 85-ton payload in its rear-dump body. Referred to as the M-85, it would be the Lectra Haul truck design that would change the mining industry forever. Shown in June 1963 is the first M-85 undergoing testing at Unit Rig's West Tulsa site, which would eventually become the company's new home for its main manufacturing plant and corporate offices.

The 85-ton capacity Lectra Haul M-85 was powered by a Cummins VT-12-700 diesel engine, capable of 700 gross/625 net horsepower. Its diesel-electric drive system utilized a GE GT-603 generator, and GE type 772 traction motors in the rear wheel assemblies, fitted with 21.00-49, 40PR-sized tires on 15x19-inch rims. Empty weight of the M-85 was listed at 103,000 pounds, with an overall length of 31 feet, on a 15-foot wheelbase. The first Lectra Haul M-85 shipped from Unit Rig in July 1963, destined for Kennecott Copper's Chino Mine, located near Silver City, New Mexico. It would soon be joined by two more M-85 units (the last truck in order leaving the factory in September 1963). The fourth truck in the initial build of four M-85 ore haulers left the factory for Anaconda's Berkeley Pit in Butte, Montana, in October 1963. Shown in August 1963 at the Chino Mine is the first M-85 built (S.N. 52).

which had been hurt by the recession. The purchase of Dart added a small number of mechanical-drive rear-dump and bottom-dump models in capacities not offered by Unit Rig. The purchase also included the large 600-series wheel loader. But just as things were looking up for the company, an unforeseen event was on the horizon that would change the future of Unit Rig and its past way of doing business.

On February 21, 1985, eight banks filed involuntary bankruptcy petitions in U.S. Bankruptcy Court in Dallas, Texas, against the Kendavis Holding Co. and Kendavis Industries International, Inc. (KIII), and 16 companies that were part of KIII, claiming they were owed $319.6 million. After 18 months of negotiations failed to reach a refinancing agreement with the Davis companies, the banks felt they had no choice but to force involuntary petitions and try to recoup their investments by liquidating the Kendavis assets. On March 14, 1985, the primary owners of the Kendavis companies, the brothers T. Cullen Davis and Ken Davis, Jr., agreed to a voluntary reorganization of their business empire under Chapter 11 proceedings, fending off the possibility of the creditors forcing immediate liquidation repayment of the past due loans. On May 20, 1985, the banks filed involuntary Chapter 11 petitions against 16 companies of Kendavis Industries (including Unit Rig), the guarantors of the loans in the original Chapter 11 filings. Even though Unit Rig was actually doing well at the time of the Chapter 11 filings, the other businesses in the Davis empire were primarily focused on the oil industry, which was in terrible shape at the time. But because Unit Rig was an asset of Kendavis, it had to conduct business as if it were also under Chapter 11. For Unit Rig, this slowed development of future products, and made the company's customers hesitate to make commitments for purchases of trucks, not knowing if the company would survive. This hurt Unit Rig, but also crippled the Dart product lines purchased the year before.

On November 6, 1986, the Kendavis companies submitted a plan for reorganization for the repayment of its debt, which was approved by the bankruptcy court (including the discharge of any remaining claims) on November 24, 1986. These rulings helped to calm the nerves of Unit Rig's customers and allowed it to conduct business in a more traditional way. Finally on May 16, 1988, the U.S. Bankruptcy Court for the Northern District of Texas granted a motion to dismiss the case against the company and allow it to emerge from Chapter 11. During the bankruptcy proceedings, Kendavis Industries agreed to sell Unit Rig as part of its restructuring plan. In October 1987, Marathon LeTourneau took an interest in Unit Rig and actually made an offer that was accepted by the company. However, the deal quickly fell apart, and another company, Northwest Engineering, ultimately reached an agreement with Unit Rig on the purchase of the firm. On June 7, 1988, Terex Corp. (the new corporate name of Northwest Engineering as of May 1988) announced that it had purchased Unit Rig & Equipment Co., renaming the company Unit Rig, Inc. On July 18, 1988, Terex made public that it had completed its acquisition of Unit Rig, including all of its U.S. and Canadian assets of the company and the stock of its foreign subsidiaries.

During the 1990s, Unit Rig would go through a series of structural realignments brought on by Terex Corp. By 1990, the company name had been shorted to that of simply Unit Rig, operating as a division of Terex Corp. In 1998 the company was made part of Terex Mining, a new division under Terex Corp., originally formed as the subsidiary Terex Mining Equipment, Inc. in the acquisition deal to purchase O&K Mining GmbH, first announced on December 15, 1997 (finalized on April 1, 1998). Unit Rig shared its standing under Terex Mining with O&K Mining, and Payhauler (also acquired in 1998).

All of the corporate shuffling of Unit Rig under Terex Corp. ownership did not seem to hinder new Lectra Haul releases by the company during the 1990s. It was during this time period that Unit Rig introduced what is perhaps one of the best mining trucks the company has ever produced—the MT-4400. Officially introduced in 1995, it stands as the company's top seller to the mining industry today, and has a production record to back it up. Most consider the M-100 Lectra Haul as the best all-around and well-balanced truck design from the early years of Unit Rig. If that is true, then the MT-4400 series of mining haulers is surely their best design in the modern era of the company.

In late 2002, Terex Mining announced that it was going to change the model designations of its Unit Rig hauler and O&K excavator product lines. The Unit Rig MT-series became the TMT-series, with the O&K RH-series becoming the TME-series. For a short time the Unit Rig mining trucks were known as the TMT 120 (formerly MT-3000), TMT 150/150AC (MT-3300/3300AC), TMT 190 (MT-3600B), TMT 205

The first four M-85 ore haulers produced a wealth of information, both positive and negative, for Unit Rig's engineers to go through. The conclusion of the company was that the design of the M-85 was sound. It only needed a few improvements overall to bring the model up to full production status. One area of redesign was the front access ladder for the operator. The original design had "cut-out" slots going up both sides of the radiator. The improved version added a ladder to the cab side and eliminated the cut-outs (which were only in the four original trucks). Shown in May 1964 is one of the first updated M-85 trucks ready for delivery to the Anaconda Company.

(MT-3700B), TMT 240 (MT-4400), TMT 260AC (MT-4400AC), TMT 360 (MT-5500), TBD 220 (BD-220), TBD 240 (BD-240), and TBD 270 (BD-270). But not all of Terex Mining management was behind the new model nomenclature changes. Management decided to change back to the "old" original model designations of both Unit Rig and O&K by June 2003, further confusing the industrial trade media and customers alike.

Unit Rig had always called Tulsa, Oklahoma, home. But market changes soon forced the company to shift its operations toward the south to take advantage of more cost-effective labor resources. On May 29, 2002, Terex mining announced that it was going to relocate its Tulsa manufacturing and assembly operations to Lubbock, Texas, with the work to be subcontracted out to Noble Construction, Inc. Along with Noble's Lubbock operations, additional work would also be handled by its Cindad Acuna, Mexico facilities. In the summer of 2004, Noble shut down its Lubbock plant and transferred all remaining Unit Rig truck related manufacturing work to its Acuna, Mexico operations. On September 7, 2004, Terex Corp. completed the acquisition of Noble CE, LLC and its Mexican subsidiary, and renamed it Terex Mexico. On February 14, 2006,

The updated Lectra Haul M-85 also received a redesign of the dump body to increase its overall payload volume. The new box design was rated at 56 cubic-yards struck and 65 heaped. The original design was rated at 47 struck and 62 heaped. Payload capacity remained at 85 tons. The new box also increased the length of the M-85 to 32 feet, 4 inches. Empty vehicle weight also increased slightly to approximately 106,000 pounds. The first six of the improved M-85 trucks were purchased by the Anaconda Company for use at its Berkeley Pit in Butte, Montana. Shown is one of three Anaconda M-85 trucks ready for shipping in May 1964. The next three trucks in the order would ship the following month in June.

The next company to place an order for the improved Lectra Haul M-85 was Kennecott Copper Corp. Kennecott was so impressed with the original M-85 ore haulers it had originally purchased in 1963, that they placed an order for an additional six trucks, also for use at its Chino Mine. The first three trucks in this order would ship from Unit Rig's Tulsa plant in July 1964, followed by the remaining three units in August. Shown in August 1964 is one of the first improved M-85 haulers to go into service at the Chino Mine.

Terex Corp. made public that they were moving the last of its management, engineering, and support personnel to Unit Rig operations now in Denison, Texas. By July of that year, most of the workers had moved on, thus bringing an end to Unit Rig as part of the Tulsa business community.

On July 1, 2003, Caterpillar, Inc. and Terex Corp. announced that they had reached a non-binding agreement in principle for Caterpillar to acquire the Unit Rig line of diesel-electric drive trucks, and for Terex to purchase Caterpillar's 5000-series mining shovel intellectual property (including the 5110B, 5130B, and 5230B). In the deal, Terex would continue to manufacture the O&K lines of hydraulic mining excavators for distribution through the Cat dealer network worldwide. But it wasn't long before an impasse was reached in the deal. After evaluating Unit Rig's records, Caterpillar did not feel the transaction was valued at what Terex wanted for it. It soon became evident that the two parties could not come to financial terms suitable to all involved in the deal. On December 10, 2003, Terex Corp. announced that it had terminated discussions with Caterpillar regarding the sale of the company's mining truck business, and that Terex would not acquire Caterpillar's mining shovel intellectual property.

Though the Caterpillar and Terex deal would end before it ever really got started, a small part of it eventually did become a reality. On August 16, 2004, Terex Corp. announced that it had reached an agreement with Caterpillar to allow the distribution and support of the Terex Mining line of O&K mining hydraulic excavators through the Cat independent dealer network. The implementation of the deal was phased into action at selected Cat dealerships worldwide throughout most of 2005.

Terex Corp. eventually found a buyer for its line of Terex mining haulers (the Unit Rig name had been dropped from use in 2008) and hydraulic excavators (O&K) that were part of Terex Mining in 2009. On December 20 of that year, Terex Corp. announced that it had signed a definitive agreement to sell its mining business to Bucyrus International, Inc. of South Milwaukee, Wisconsin. Along with the diesel-electric trucks and the hydraulic excavators, the deal would also include Terex's track and rotary blasthole drills (Reedrill) and highwall miner (Superior) equipment lines. On February 19, 2010, Terex Corp. and Bucyrus completed the transaction. Now the old Unit Rig line of diesel-electric drive haulers would be referred to as "Bucyrus Mining Truck," if only for a brief time.

Without a doubt, the biggest news story in the heavy equipment manufacturing industry in 2010 was the announcement on November 15 of that year that Caterpillar, Inc. and Bucyrus International had entered into an agreement under which Caterpillar would acquire Bucyrus International. Included in the deal were all of the previous Terex Mining product lines purchased earlier in the year, along with the company's rope shovel, dragline, blasthole drill, and underground mining equipment lines. After Bucyrus stockholders' approval (January 20, 2011), and clearance by the United States Department of Justice (DOJ, May 20, 2011), Caterpillar would complete its acquisition of Bucyrus on July 8, 2011.

One of the first things Caterpillar would do after the completion of the deal was to "retire" the Bucyrus name from all of the newly acquired product lines, and reinstate the old "Unit Rig" brand name back onto the diesel-electric drive truck line. At this time, it is hard to see what the future holds for Unit Rig branded truck lines of Caterpillar, Inc. There are many questions that need to be resolved over the coming months as to how the AC-drive Unit Rig lines will be marketed alongside Cat's mechanical-drive and electric mining truck offerings. At the time of this writing, Caterpillar is going through a process of integrating more of its compo-

When the Lectra Haul M-85 was first released, it was equipped with 21.00-49, 40PR-sized tires. In 1965 optional larger 24.00-49, 42PR tires on 17x49-inch rims were offered for the first time. The first twelve M-85 trucks to be equipped with these larger tires shipped from the factory between May and July 1965 to Molybdenum Corporation of America (Molycorp), located near Questa, New Mexico. Pictured in July 1967 is one of these M-85 trucks at Molycorp.

nents into the Unit Rig designs, such as engines, hydraulics, electrics, and drive systems. Also, the trucks will continue to be assembled in Acuna, Mexico, for the foreseeable future. But one has to wonder just how long Caterpillar will retain Unit Rig as a "product line" or survive as individual truck lines. Only time will tell.

The Lectra Haul M-85 sold in large numbers and was very successful (and profitable) for Unit Rig, as well as the mining companies that operated them. Shown around 1980 is the last version of the M-85 built, featuring the updated operator's cab that was first introduced in 1977, just before delivery to a mining operation in Mexico. This last version of the M-85 featured 24.00-49-sized tires as standard, and had a listed empty vehicle weight of 117,150 pounds.

As truck sizes increased in capacity in the mid-1960s, there became a clear need for more powerful diesel engines. But the engine manufacturers were slow to develop larger powerplants without an established customer base in the industry. Unit Rig was the first heavy equipment manufacturer to design and build prototype mining haul trucks powered by high-horsepower gas turbine engines. The first model built was the Lectra Haul M-100 Turbine Ore Hauler equipped with a 1,100 gross horsepower IH Solar turbine engine. This truck was the first M-100 produced (S.N. 120) and it would ship in October 1965 to Anaconda's Berkeley Pit for field trial testing (Anaconda fleet #J626). The second M-100 turbine equipped unit was fitted with a 1,200 gross horsepower General Electric LM-100 gas turbine engine. This truck (S.N. 121) was shipped from Unit Rig in September 1965 to Las Vegas for displayed at the October American Mining Congress show (pictured).

This detail view of the Lectra Haul M-100 Turbine Ore Hauler that was displayed at the AMC show in October 1965 shows the large exhaust of the General Electric LM100 gas turbine engine. The lightweight, 1,200 horsepower GE turbine powerplant weighed in at only 350 pounds, which was considerably lighter than a diesel engine. Drawbacks to the turbine design were its overall initial cost, and a fuel usage that was more than double that of a diesel. After the AMC show, the GE gas turbine-engined M-100 was shipped to Kennecott Copper's Chino Mine for further testing.

The third (and last) M-100 gas turbine truck built by Unit Rig was equipped with the 1,100 horsepower IH Solar powerplant, just like the first M-100 produced. The easiest way to tell the Solar turbine equipped M-100 from the GE version were the exhausts. The GE's exited on the right side, behind the right front tire, while in the Solar design the exhaust was a massive stack protruding from the top of the engine hood. Shown at Unit Rig's Tulsa plant in July 1967 is the third M-100 gas turbine equipped truck featuring the IH Solar engine, destined for Kennecott Copper Corporation's Bingham Canyon Mine, located just outside Salt Lake City, Utah. Note the new design of the ladder on the front of the truck, which was introduced into production in 1967 for both the M-100, as well as the M-85 Lectra Haul model lines.

Unit Rig first started to ship diesel engine powered M-100 Lectra Haul trucks in January 1966. These early units were fitted with a Cummins VTA-1710-C diesel, rated at 800 gross/725 net horsepower. The electric drive system utilized a GE GT-603 generator and GE 772 wheel motors. The M-100 utilized the same 15-foot wheelbase frame design as the M-85, but reinforced in key areas to handle the trucks increased 100-ton payload capacity. It also was equipped with the same DYNAFLOAT rubber cushioned suspension system as the M-85. Empty weight of the early M-100 was approximately 110,000 pounds. Standard tire size was 24.00-49, 42PR. Shown in December 1966 is one M-100 from a big order of 25 trucks purchased by Cyprus Pima Mining Company for its Pima Mine, located south of Tucson, Arizona.

The Lectra Haul M-100 proved to be just as popular in the mining industry as the M-85. In 1967 the model line received a few upgrades, as well as expanded engine option choices. The biggest visual change was the redesign of the front ladder system, which was a much safer design than the previous vertical fabrication. Along with the Cummins Diesel, customers could now specify a Detroit Diesel 12V-149NA (800 gross/725 net horsepower), or a 12V-149T (1,000 gross/920 net horsepower) version. Picture taken in 1967.

As the Lectra Haul M-100 matured in the product line, its overall empty weight also increased to approximately 131,500 pounds, but its payload capacity remained at 100 tons. An optional 27.00-49, 42PR tire size was also now offered. Image date is 1967.

Unit Rig first started to experiment with electric trolley assist power systems in 1967 when it modified one of its Lectra Haul M-100 trucks for testing at Kennecott Copper's Chino Mine in New Mexico. The pantograph-equipped test M-100 was powered by a 700 gross/625 net horsepower Detroit Diesel 16V-71T engine, utilizing GE 772 electric wheel motors. During testing, the M-100 was capable of hauling a 123-ton payload up a 7-percent incline, at a speed of just over 13-miles-per-hour under trolley assist. The success of these tests led a few of Unit Rigs customers to convert their fleets of Lectra Haul trucks over to trolley assist systems. In the right applications, the trolley system improved energy consumption and increased power on grades for higher top speeds. Picture taken at the Chino Mine in 1967.

Towards the end of the 1970s, the M-100 was offered with still more powerful engine choices, as well as an all-new cab design first introduced in 1977. New powerplant options included two Cummins KTA-2300-C diesels (1,050 gross/950 net horsepower, and 1,200 gross/1,075 net horsepower), and a more powerful Detroit Diesel 12V-149TI (1,200 gross/1,070 net horsepower). The previous Detroit Diesel 12V-149T was also still offered for the M-100. The optional 27.00-49 tires now became standard size of choice throughout the 1970s. Shown in October 1979 is an updated M-100 Lectra Haul operating in South America at CVG—Ferrominera Orinoco's El Pao Mine in Venezuela.

At the 1974 AMC show in Las Vegas, Unit Rig displayed a new concept in driver-less mining trucks called Automatic Truck Control, better known as ATC. Developed in cooperation with Saab-Scania, ATC was an electronic system utilizing a master control unit that programmed the truck's fail-safe steering, direction control, truck separation, speed selection, and braking, including emergency and equipment malfunction events. A guide-wire buried in the path of the truck would allow the unit to track along a predetermined path, through a series of electronic sensors mounted on a bracket just below the front bumper. Pictured in April 1980 is a test M-100 Lectra Haul.

Unit Rig tested its automated ATC driver-less system on an M-100 truck at the Tulsa plant from 1976 to 1977. Once changes were made to the prototype system, a series of additional tests were performed on five used M-100 Lectra Hauls equipped ATC at Kennecott Copper's Chino Mine starting in 1978. These tests would continue on until mid-1980, at which time the ATC project was abandoned due to insurmountable technical difficulties, and a downturn in spending in the mining industry caused by the looming worldwide economic recession. Shown in April 1980 is one of the white M-100 Lectra Haul ATC equipped trucks at the Chino Mine.

Unit Rig's original M-120 Lectra Haul was essentially an M-85/M-100 equipped with a more powerful engine, larger dump body, and bigger tires. The M-120 utilized the same basic frame design (as well as 15-foot wheelbase) as its two smaller payload brothers, with areas strengthened to accommodate the trucks greater rated payload capacity of 120 tons. Standard engine was the Detroit Diesel 12V-149TI, rated at 1,200 gross/1,070 net horsepower. Standard tire size was 27.00-49 series type. The first five M-120 units would ship from Unit Rig in March 1968 to Iron Ore Company of Canada's Schefferville, Quebec, mining operation. In 1974, an additional M-120.17 Lectra Haul model was added to the product line featuring a 17-foot wheelbase. Unit Rig would rename all previously delivered M-120 trucks as M-120.15 Lectra Hauls to reflect that models use of a 15-foot wheelbase. Pictured in 1977 is a larger model M-120.17 operating at Petrotomics uranium mine (a division of Getty Oil), in Shirley Basin, Wyoming.

The M-120.17 Lectra Haul introduced in 1974 was actually the Mk 30 from 1971 with a new model identification (see Mk 30). Unit Rig officials felt that the original Mk 30 nomenclature should be saved for use later on a new truck design, and not one based on the lengthened frame design of the M-120.15 with an extra two feet added to the wheelbase. The first four M-120.17 Lectra Hauls assembled would ship to AMAX Coal's Belle Ayr Mine, located just south of Gillette, Wyoming, in April 1974. Shown in 1977 is an M-120.17 working at Larco, Societe Miniere et Metallugique de Larymna, S.A. nickel mine, located near Athens, Greece.

The frame of the M-120.17 Lectra Haul was based on the design of the M-120.15, but stretched to a 17-foot wheelbase so the truck could accept larger 30.00-51, 46PR-sized tires. Standard engine choice was the Detroit Diesel 12V-149TI (1,200 gross/1,070 net horsepower). By 1977 a new Cummins KTA-2300-C (1,200 gross/1,075 net horsepower) was also added to the list of available engine choices. The electric drive system utilized a GE GT-603 generator and type 772 traction wheel motors. Payload capacity was 120 tons. Empty weight was approximately 160,800 pounds, with an overall length of 34 feet, 9 inches. In comparison, the M-120.15 empty weight was 145,000 pounds, with a length of 32 feet, 9 inches. Shown is an M-120.17 taken in late 1979.

In April 1978 Unit Rig shipped a special ordered M-120.17 Lectra Haul truck (S.N. 1944) to Goodyear Tire & Rubber Company's Tire Testing Division in San Angelo, Texas. This unit, along with the previously shipped Unit Rig M-200 from 1972, was used to test the performance of various tire designs produced by the company for use in off-highway haul truck applications. Picture taken in mid-1978.

In 1966 Kaiser Steel presented Unit Rig with a proposal for the design of a 200-ton capacity, rear-dump mining truck for use at its expanding coal mining operations located near Sparwood, British Columbia, Canada. Unit Rig accepted the challenge and over the next two years designed and built the first prototype truck to meet Kaiser's specifications, the Lectra Haul M-200. Shown in October 1968 at Caesars Palace in Las Vegas, was the star of the American Mining Congress show, the pilot Lectra Haul M-200 (S.N. 51).

The Lectra Haul M-200 was designed around a diesel-electric drivetrain supplied by GM's Electro-Motive Division (EMD). The frame of the M-200 utilized a 22-foot wheelbase, giving the truck an overall length of 43 feet, 4 inches, and a width of 24 feet. Specified tires were the largest available at the time, 36.00-51, 58PR series type. But the chassis layout of the hauler was designed to accept far larger tires as they became available to the mining industry. After the mining show in Las Vegas, the pilot M-200 was eventually shipped to Cyprus' Pima Mine, located just south of Tucson, Arizona, for engineering field evaluation tests before being shipped up to its final home in British Columbia. Pictured in early 1969 at the Pima Mine is the prototype M-200. Note the redesign of the exhaust system as compared to the structure utilized on it at the mining show.

Unit Rig would ship its first seven M-200 trucks (including the prototype unit at Pima) up to Kaiser Resources Ltd. (Canadian subsidiary of Kaiser Steel) starting in the first quarter of 1969, with the first trucks assembled on site in April of that year. The rest of the initial shipment of trucks would leave the factory in September. In all, Kaiser would operate 22 of these giants, which for a time were the world's largest two-axle, rear-dump trucks. Empty weight of the M-200 was approximately 264,500 pounds. Payload capacity was a big 200 tons, which was considered huge in its day. Shown at work in 1969 is one of the Kaiser M-200 haulers being loaded by a 25-cubic-yard P&H 2800 electric mining shovel.

The early Lectra Haul M-200 trucks were powered by a huge 5,160-cubic-inch, 8-cylinder EMD 8-645-E4 locomotive diesel engine, rated at 1,650 gross/1,500 net horsepower, at just 900 rpm. Its major electric drive components consisted of an EMD-AR5 alternator and two flange-mounted EMD-79UR traction wheel motors in the rear. These were connected to newly-designed Unit Rig W-200 double planetary drive gear wheel assemblies. The M-200 also utilized the company's DYNAFLOAT rubber cushioned, column-type suspension system. Pictured is the 8-cylinder EMD engine module for the Lectra Haul M-200.

Severe wear on the early 36.00-51-sized tires led Unit Rig to adopt the larger 40.00-57, 60PR series type, mounted on bigger 57-inch rims, as soon as they were made available in early 1971. This raised the truck's empty weight to approximately 284,600 pounds (later increased to 299,200 pounds). In March 1975 Unit Rig placed a special Tulsa-built, 12-cylinder engine equipped M-200 in a coal mine in Mezhdurechensk, East Siberia (then part of the U.S.S.R.). The success of this test truck led to the placement of a huge order of M-200 Lectra Hauls for the U.S.S.R. totaling 84 units (contract for 30 units awarded in April 1976, with the remaining 54 awarded in December 1977). All 84 trucks were assembled at Unit Rig's new facilities in Stevensville, Ontario, located 73 miles south of Niagara Falls. To cope with the extreme working conditions in Siberia, the trucks were fitted with more powerful 12-cylinder EMD 12-645-E4 locomotive engines, rated at 2,450 gross/2,250 net horsepower. Capacity remained unchanged at 200 tons, with an empty weight of 316,600 pounds. Length was 48 feet, with a width of 25 feet, 6 inches. The first three trucks of the order arrived on site in Neryungri, Western Siberia, in July 1977, with the last trucks of the order shipping in October 1980. Shown is one of the Russian M-200 trucks in 1977 just before shipping. Note the longer nose of the unit, as well as the four sections of the radiator grille, indicating the installation of the larger 12-cylinder engine.

All of the Lectra Haul M-200 trucks built by Unit Rig over its entire production run were of a rear-dump variety, except one. In 1972 the company designed a special M-200 test truck for Goodyear Tire & Rubber Company's Tire Testing Division, for use at its proving grounds in San Angelo, Texas. The tractive-effort test rig was completed and shipped to Goodyear in October 1972. Total number of all M-200 Lectra Hauls produced was 120 units. Pictured in mid-1978 is the one-off M-200 Lectra Haul test truck (S.N. 79).

In 1981 Unit Rig released a new version of its older tried and true M-85 in the form of the Mk 24 Lectra Haul. Still rated with an 85-ton payload, the Mk 24 featured improvements to the frame, as well as an all-new front radiator design. Wheelbase remained unchanged at 15 feet. The Mk 24 was offered with three standard 12-cylinder engine choices: the Cummins VTA-1710-C (800 gross horsepower), the Cummins KT-2300-C (900 gross horsepower), and the Caterpillar D-348 (870 gross horsepower). The electric drive system was all GE, consisting of a GTA-23 alternator and two 772-XS1 traction wheel motors. Shown in April 1981 is the pilot Mk 24 Lectra Haul.

The Unit Rig Mk 24 Lectra Haul rode on 24.00-49, 48PR-sized tires, and had an empty vehicle weight of approximately 133,667 pounds. Overall length of the truck model was 33 feet, 9 inches. The Mk 24 was a slow seller in the product line, and the worldwide recession of the early 1980s did little to improve its sales outlook. In 1994, the Mk 24 was upgraded into the 100-ton capacity Mk 27, but the mining and quarry marketplace showed little interest in the diesel-electric drive 100-ton capacity off-highway truck. Shown in November 1981 is the prototype Mk 24 Lectra Haul in service at Triple Elkhorn Mining Company's operation in Kentucky.

To satisfy marketplace demands for 120-ton capacity mining trucks fitted with larger 30.00-51, 46PR tires, Unit Rig took the frame of the 15-foot wheelbase M-120 and added two feet to the wheelbase of it, enabling it to accept the larger tire size. Referred to as the Mk 30 Lectra Haul, its nomenclature reflected the tire size and not the capacity as in past releases. Drivetrain specifications were virtually identical to the previous mentioned M-120.17. The pilot Mk-30 shipped to Cyprus' Pima Mine for long term testing in September 1971. The next 15 trucks would all be ordered by Iron Ore of Canada (I.O.C.) for use at its Carol Lake operation in Labrador City, (Labrador) Newfoundland, Canada. The first 13 trucks would all ship by April 1972, followed by two more units in January 1973. After delivery of the first 16 trucks, Unit Rig decided to change the trucks nomenclature back to the old system based on the capacity, updated with the wheelbase size. When the next 17-foot wheelbase Lectra Haul shipped in April 1974, it would be known as the M-120.17. It was at this time that the 15-foot wheelbase M-120 was reclassified as the M-120.15. Pictured in September 1971 is the pilot Mk 30 Lectra Haul (S.N. 1015).

The Mk 30 nomenclature would once again be dusted off for use on a special order of trucks built for Inland Steel Mining Company's Minorca Mine, located in Virginia, Minnesota. This version of the Mk 30 was rated with a 130-ton payload capacity, and was fitted with a 1,200 gross horsepower diesel engine. Tire specified was the 30.00-51, 46PR, but a larger 33.00-51 could be fitted as an option to meet payload requirements. Empty weight of this Mk 30 was 156,061 pounds. A total of 15 of these 130-ton capacity spec'd Mk 30 trucks were delivered to Inland Steel between July 1976 and February 1979. Pictured in March 1979 is one of the 130-ton capacity Mk 30s at the Minorca iron ore mine.

In early 1981 Unit Rig introduced a new Mk 30 Lectra Haul that would now finally become a permanent model in the product line. Because of frame problems with the previous designs, Unit Rig engineers updated many areas of the 17-foot wheelbase chassis to make the new Mk 30 more reliable in the field. Another change was the lowering of the payload capacity to 120 tons. The first of the improved Mk 30 trucks (S.N. 51) was shipped to Inland Steel's Minorca Mine in February 1981. Pictured in January 1981 is one of the updated Mk 30 Lectra Hauls. Note the slight redesign of the radiator grille front as compared to earlier versions of the Mk 30.

The improved Mk 30 Lectra Haul was offered with three engine choices initially. They included the Detroit Diesel 12V-149TI, the Cummins KTA-2300-C, and the Caterpillar 3512, all rated at 1,200 gross horsepower. By 1986 the Cummins KTA-38-C replaced the earlier Cummins offering. The electric drive system utilized a GE GTA-25 alternator and GE 772-YS traction wheel motors. Optional heavy-duty GE 776-HS motors were also offered for deep pit applications. Standard tire choice was the 30.00-51, 46PR series type. Empty weight of this model was 170,750 pounds, with an overall length of 34 feet, 10.5 inches. Shown is a Canadian-built Mk 30 operating at a copper mine in Quebec around 1982.

The Unit Rig Mk 36 Lectra Haul was without a doubt one of the most popular 170-ton payload class diesel-electric drive mining trucks ever offered to the industry. Its classic looks and overall proportions were unmistakable and would become the face of Unit Rig Lectra Haul marketing for years to come. Unit Rig first introduced the Mk 36 in 1972, shipping the first unit to Cyprus' Pima Mine in August of that year (S.N. 51). Shown at the Pima Mine in early 1977 is one of the first Mk 36 Lectra Hauls to feature the new operator's cab design with the larger glass areas (S.N. 285). This unit shipped from the Tulsa factory to Pima in March 1977. It would later be transferred to the Anamax Twin Buttes Mine in Sahuarita, Arizona, in June 1978.

The Mk 36 Lectra Haul utilized a frame design incorporating a 17-foot, 6-inch wheelbase. The Mk 36 shared this same chassis with its smaller brother in the product line, the 150-ton capacity Mk 33. In fact the Mk 33 was nothing more than the Mk 36 equipped with 33.00-51, 46PR tires and an engine de-rated to 1,350 gross horsepower. Few Mk 33 trucks were built in the 1970s, with the first three delivered in October 1972 (S.N. 52-54) to the I.O.C. Sept-Iles terminal docks in Quebec, Canada (later, these three same trucks would be rebuilt into full Mk 36 configurations). The next two lots of Mk 33s would ship from Unit Rig (six in April 1974, and eight in November 1976) to the McIntyre Mines in Alberta. Factory records indicate that by the end of 1981, that was the total number of Mk 33 Lectra Hauls delivered by the company. Shown in July 1978 is a Mk 36 operating at Anaconda's Berkeley Pit in Butte, Montana.

In August 1979, Unit Rig produced its 2,500th Lectra Haul mining truck. The special white Mk 36 is shown here rolling off Unit Rig's Tulsa assembly line during a small ceremony to celebrate the company's great production milestone. This Mk 36 would then ship to Kennecott Copper's Chino Mine in New Mexico, which was also the home for the company's first delivery of a production M-85 Lectra Haul back in July 1963.

The Mk 36 Lectra Haul was offered with two 16-cylinder engine choices: a Detroit Diesel 16V-149TI or a Cummins KTA-3067-C, both rated at 1,600 gross/1,450 net horsepower. The electric drive system featured a GE GTA-22 alternator and GE 776 traction motors mounted in the rear wheel assemblies. This Mk 36 is part of Erie Mining Company's fleet of Lectra Haul trucks operating at its Hoyt Lakes iron mining operation in Minnesota, in May 1980.

The "36" in the nomenclature of the Mk 36 represented the standard tire size fitted to the truck, in this case 36.00-51, 50PR series type. Maximum payload capacity of the Mk 36 was 170 tons. Its empty vehicle weight in the early 1970s was approximately 194,825 pounds. By 1982 this figure had risen to 213,066 pounds (220,276 pounds with full liquids, half fuel load, and operator). Length of the model was 38 feet, 9 inches, with a width of 23 feet, 3 inches. Picture taken in May 1980 at Erie Mining.

Despite some early frame problems with the Mk 36 in the field, the model went on to become a top seller for Unit Rig with a reported number of rear-dump units shipped into service of approximately 685 (524 units by the end of 1981 alone) by the time the model was removed from the product line in 1990. Shown is a Mk 36 taken around 1982. Not the use of the new Unit Rig logo on the upper side of the radiator. This new design would begin to appear on Lectra Haul trucks starting in late 1981.

Unit Rig offered a trolley assist upgrade package for the Mk 36 Lectra Haul, which proved quite popular in mining operations set up for this type of haulage application, especially in South Africa. The conversion could be performed on new or used trucks, and consisted of a set of front-mounted pantographs that drew power from overhead lines, driving the electric traction wheel motors. This set-up increased fuel economy and allowed the trucks to obtain faster speeds on grade, making them ideal for deep pit mining operations. The image of a trolley equipped Mk 36 was taken at Rio Tinto's Palabora copper mine in the Republic of South Africa, in the early 1980s.

The early frame failures of the M-120.17, Mk 30, and Mk 36, led Unit Rig engineers to fabricate a totally new design, not based on a stretched version of its original 15-foot wheelbase M-85/100/120 chassis. Referred to as the MT-1900 Lectra Haul, it was designed around a 19-foot wheelbase, and was rated as a 190-ton capacity hauler, capable of running on 36.00-51, 58PR-sized tires. Engines offered for this design were the Cummins KTTA-50-C1800, and the Detroit Diesel 16V-149TIB, both rated at 1,800 gross/1,704 net horsepower. The GE electric drive system consisted of a GTA-22D alternator and powerful 788 traction wheel motors. Empty weight of the truck was 247,900 pounds. Overall length was 41 feet, 2 inches, with a width of 23 feet, 6 inches. Only two MT-1900 trucks were indicated as being built. The first unit was to ship to Utah International Island Copper in British Columbia, Canada, the second would go to RTB in Yugoslavia. The trucks proved very problematic in the field and a complete re-evaluation of the design would be required to salvage the program. Shown in mid-1984 is the pilot MT-1900 Lectra Haul.

The next step in the evolution of the MT-1900 program was the MT-2050 Lectra Haul. Introduced in mid-1986, the MT-2050 took what was good with the MT-1900 design, and improved the areas that were not. The MT-2050 was capable of carrying a nominal 205-ton payload made possible by the use of larger 37.00-57 (E-4) tires on 57-inch rims. Powertrain choices were the same as the MT-1900 with the addition of an optional Caterpillar 3516 diesel, rated at 2,000 gross horsepower. Empty weight of the MT-2050 was listed at 268,800 pounds. Overall length of the truck was 42 feet, 1 inch. Only a handful of these Lectra Hauls were placed into service. Shown in October 1995 is a fully loaded MT-2050 in operation at the Cyprus-AMAX Belle Ayr Mine, located just south of Gillette, Wyoming. *Image by the author*

The smallest of the modern MT-series of Lectra Haul truck designs was the limited production MT-2700. First announced by Unit Rig in late-1995 as a replacement in the product line for the Mk 27, the MT-2700 Lectra Haul was equipped with 30.00-51, 46PR-sized tires on 51 inch rims, utilizing a fully modern nitrogen over oil strut suspension system. Engine choices were either the Detroit Diesel 12V-149TIB DDEC III, or the Cummins KTA-38E (both de-rated to 1,050 gross horsepower). The drive system utilized a GE GTA-25 alternator and GE type 776 wheel motors. Capacity of the MT-2700 was a nominal 100 tons. Pictured in June 1999 is one of five MT-2700 Lectra Hauls originally purchased for use at CCB's Guarain-Ramecroix limestone quarry in Belgium. Two of these trucks were later moved to the Carrieres d'Antoing quarry, also in Belgium. *Urs Peyer*

Introduced in 1996, the 120-ton capacity Unit Rig MT-3000 Lectra Haul would replace the aging Mk 30 in the product line. Both the MT-3000 and MT-2700 shared the same 17-foot, 6-inch wheelbase chassis and frame, as well as the same standard-sized tires. Both also utilized the identical Detroit Diesel and Cummins engine offerings, but with higher power ratings for the MT-3000 of 1,200 gross/1,125 net horsepower for the Detroit, and 1,200 gross/1,129 net horsepower for the Cummins. Empty weight of the truck was listed at 214,880 pounds, with an overall length of 39 feet, 11 inches. The MT-3000 Lectra Haul is shown here at its world unveiling at the 1996 MINExpo show in Las Vegas, Nevada, held in September of that year. *Image by the author*

Unit Rig first announced its 150-ton capacity MT-3300 Lectra Haul in 1994 as the replacement for the Mk 33 in the product line. The original MT-3300 utilized a DC electric drive system by GE. In the year 2000 the company introduced an AC-drive version of the MT-3300. The MT-3300AC utilized the new GE150AC Drive System consisting of a GTA-22 alternator, IGBT controls, and GEB23 traction wheel motors. Standard engine for the model was the 12-cylinder Detroit Diesel 12V4000, rated at a maximum 1,875 gross/1,720 net horsepower. The truck design utilized a 17-foot, 6-inch wheelbase, and was fitted with 33.00 R51 radials as standard equipment. Empty vehicle weight was approximately 232,000 pounds, with an overall length of 39 feet, 11 inches. The first four MT-3300 AC trucks built would ship to an Imerys' member magnesium mining operation in Alabama. By 2011 approximately 147 had been placed into service worldwide. Shown is an MT-3300AC hauler on display at MINExpo in Las Vegas, in September 2000. Note the absence of the "Lectra Haul" trade name, which had been discontinued from use by the end of 1999. *Urs Peyer*

The introduction of the MT-3700/3600 Lectra Hauls in 1990 was a game changer for Unit Rig in many ways engineering-wise. The trucks featured an all new frame design incorporating a new drop box-beam front axle design made partially possible by utilizing the technology acquired through the purchase of Dart in 1984. The MT-3600 was virtually identical to the MT-3700 except for the use of less powerful wheel motors (GE 788 for the MT-3700, and GE 776 type for the MT-3600) and smaller tires and wheels. Tires for the MT-3700 were 37.00- R57 type on 57-inch rims, while the MT-3600 made due with 36.00-51 tires mounted on 51-inch rims. The MT-3700 was rated as a 190- to 205-ton capacity hauler, while the MT-3600 carried a slightly lower payload range of 170 to 190 tons. Wheelbase of the MT-3700 was 18 feet, 6 inches, with the MT-3600 measuring in at 18 feet, 2 inches. Overall length of both trucks was 41 feet, 10 inches, with a maximum width of 24 feet, 3 inches. Pictured is the pilot MT-3700 in 1990.

The MT-3700/3600 Lectra Hauls both utilized the same standard powerplant choices of 16-cylinder engines—the Detroit Diesel 16V-149TIB or the Cummins KTA/KTTA-50-C. Both engine packages were available in different horsepower settings to meet customer's specific needs. Power ratings of 1,600 gross/1,454 net horsepower, 1,800/1,654, and 2,000/1,854 were available for both engines as installed in the MT-3700. The MT-3600 was not offered with the 2,000 horsepower option. Both models also utilized fully modern nitrogen-over-oil strut suspension systems. Empty weight of the MT-3700 was listed at 266,517 pounds (231,643 pounds for the MT-3600). Updated "B" versions of both models were introduced in 1996. Pictured is an MT-3700 Lectra Haul on display at the 1992 MINExpo in Las Vegas, Nevada. *Author's collection*

In 2005 Unit Rig introduced an AC-drive version of the model identified as the MT-3700AC. The new truck model featured a GE IGBT AC-drive system consisting of a GTA-41 alternator and two GEB25B traction wheel motors. Top engine choices were the MTU/DDC 16V4000 and the Cummins QSK60. Both of these engines had identical power outputs with two power option set-ups—2,300 gross/2,150 net horsepower, and 2,500 gross/2,350 net horsepower. Empty vehicle weight is approximately 330,000 pounds, with a payload capacity of 205 tons. Standard tire choice was 40 R57 radials, with an option for larger low-profile 46/90 R57 series type. Wheelbase is 18 feet, 6 inches, with an overall length of 42 feet, 10 inches, and a width of 24 feet. The MT-3700AC holds the honor of being the first trucks delivered into service in the Bucyrus color scheme, to the New Gold Mesquite Mine in Southern California, in May 2010. In July 2011 the model once again became a "Unit Rig" (now simply referred to as a MT-3700) after the purchase of Bucyrus by Caterpillar. Number of MT-3700AC units sold by 2011 was approximately 255 units. Picture of MT-3700AC taken in March 2006. *Terex*

The Unit Rig MT-2050 was only offered for a short time as the company's engineers figured out ways to get more capacity out of the truck's 19-foot wheelbase frame design. By increasing power and fitting larger 40.00-57, 60PR tires, Unit Rig was able to increase the new model's payload capacity to a nominal 212 tons, with a job rated range of 210 to 240 tons maximum. Referred to as the MT-2120 Lectra Haul, the first truck in the series would ship to Rio Tinto's subsidiary Hamersley Iron's (now Pilbara Iron) Paraburdoo Mine in Western Australia, as part of a 21 truck order for the MT-2120. The first MT-2120 is shown here in January 1987 at a dedication ceremony for the new hauler just before being placed into service at Paraburdoo.

Standard engine offered for the MT-2120 Lectra Haul was the Detroit Diesel 16V-149TIB (optional 2,200 gross/2,054 net horsepower), with an optional Cummins KTTA-50-C available. Standard power output for both engines was 2,000 gross/1,854 net horsepower. The main electric drive system consisted of a GE GTA-26 alternator and GE 787 motorized wheels with integral dynamic retarders. Empty weight of the truck was approximately 301,365 pounds, with an overall length of 43 feet, 2 inches, and a maximum width of 24 feet, 6 inches. Unit Rig offered a nitrogen-over-oil strut suspension system for the front wheels only. The rear had to make do with the old Unit Rig column type rubber suspension design. Shown in February 1987 is one of the MT-2120 Lectra Hauls operating at the Paraburdoo iron mine in Western Australia. These first trucks were equipped with the standard 2,000 gross horsepower Detroit Diesel engine.

In 1988 Unit Rig changed the model name of the MT-2120 to that of the MT-4000, to reflect the tire size utilized on it, and not the nominal tonnage rating. The MT-4000 Lectra Haul was well received in the marketplace and was considered a sales success for Unit Rig. It now seemed that the company's persistent frame problems were finally a thing of the past. Image taken at the Cyprus-AMAX Belle Ayr Mine in October 1995. *Image by the author*

Lectra Haul mining trucks were always popular with customers operating in the Powder River Basin coal region of Wyoming, and the MT-4000 was no exception. Shown in October 1995 at Cyprus-AMAX's Belle Ayr Mine is an early MT-4000 equipped with a huge PHIL Hivol coal body provided by Philippi-Hagenbuch, Inc. These high-volume rear-dump bodies are perfectly suited for hauling lighter weight coal payloads, especially when equipped with a tailgate such as this one is. Though the volume of material is massive, the truck's payload is still maintained at 240 tons. It is truck and dump body combinations such as this that has limited the sales of tractor-trailer bottom-dump coal haulers in recent times. *Image by the author*

In 1990 Unit Rig introduced a series of upgrade refinements to the MT-4000 Lectra Haul product line. The new MT-4000 still offered the same standard engine choices found in the early version of the truck, but now offered an additional 20-cylinder Detroit Diesel 20V-149TIB, rated at 2,500 gross/2,334 net horsepower. The truck also utilized a nitrogen-over-oil strut suspension system on all four corners, with the rear units featuring two-stage spring rate and two-stage damping features, a giant improvement over the old rubber cushion designs from the past. Other improvements were its innovative staggered-step access ladder on the front of the unit, which was considered one of the safest designs in the industry, and more durable 40.00-57, 68PR tires. Pictured in 1990 is one of the improved MT-4000 Lectra Hauls.

The improved MT-4000 was Unit Rig's direct counter attack in its product line to Dresser's impressive 830E Haulpak, since both trucks competed for the same potential 240-ton capacity, diesel-electric drive mining truck customers. The MT-4000 was now rated as a true 240-ton capacity hauler. Its empty vehicle weight was approximately 309,268 pounds. The overall length was 43 feet, 2 inches, with a width of 25 feet, 8 inches, on a 19-foot wheelbase. Image taken in 1990.

In the early 1990s Unit Rig engineers developed a special cooling package for the MT-4000 consisting of radiators having two parallel fans. This "super-cooling" option extended the front end of the truck to accommodate the massive cooling package, giving them a very awkward, nose-heavy appearance. The first MT-4000 Lectra Hauls to receive this option were built for use at Boliden-Apirsa's Aznalcollar open-pit zinc mine, located west of Seville, in southern Spain, in 1992 (pictured). The sixteen MT-4000 trucks delivered to the mine regularly operated in high ambient temperatures, often exceeding 120 degrees Fahrenheit. The MT-4000 fleet was equipped with Detroit Diesel 16V-149TIB engines, rated at 2,200 gross/2,054 net horsepower. One of the trucks in the order was specified with the larger 20-cylinder 20V-149TIB engine, rated at 2,500/2,334 horsepower. *Terex*

The Unit Rig MT-4400 Lectra Haul was one of the company's finest modern mining truck designs, with some saying it was the best truck they *ever* produced. There is no denying that the MT-4400 had a profound effect on the product line and would influence all future Unit Rig truck production. Unit Rig first released specifications on the new truck in December 1994. One of the standout elements of the hauler was its new clear-vision operator's cab design, featuring a curved, wrap-around windshield, a first for a mining truck in the industry. This design theme carried on through other areas of the truck as well, such as the curved sides of the radiator housing sheet-metal. But the MT-4400 was more than just a pretty face. Production was what this truck was all about, and a nominal 260-ton payload made many in the industry sit up and take notice. Shown in May 1995 at the Tulsa plant was the first MT-4400 completed.

The standard engine in the early MT-4400 Lectra Hauls was the 16-cylinder MTU/DDC 16V396TE diesel, rated at 2,467 gross/2,287 net horsepower. Customers could also request the Cummins K2000E (2,000 gross/1,854 net horsepower) as an alternate choice. By 1998 two new available engines replaced the original two offerings; the MTU/DDC 16V4000, and the Cummins QSK60, both rated at 2,500 gross/2,287 net horsepower. The DC-electric drive system for all of these engines utilized a GE GTA-26 alternator and GE 787 traction wheel motors. The first fleet of MT-4400 Lectra Hauls was delivered to Powder River Coal Company's Caballo Mine, located just south of Gillette, Wyoming, in the summer of 1995, and all were equipped with the MTU/DDC 16V-396TE engine package. Shown is one of Caballo Mine's MT-4400 trucks in October 1995. *Image by the author*

The MT-4400 was designed around a very robust frame design, featuring a 21-foot wheelbase. The big Lectra Haul also featured Unit Rig's box-beam front axle design first seen in 1990 on the MT-3700/3600 model series. The hauler also utilized the company's exclusive staggered-step front ladder design. Empty vehicle weight of the early MT-4400 was approximately 344,572 pounds. Overall length was listed at 45 feet, 7 inches, with a width of 24 feet, 10 inches. Image taken in October 1995. *Image by the author*

The Unit Rig MT-4400 Lectra Haul was originally designed around the use of a new 44.00 R57 tire (relating to the "44" in the trucks nomenclature) that was supposed to be available in 1995, with a targeted nominal payload rating of a very optimistic 280 tons. By the time the MT-4400 was ready for shipping, the only tire available for it was the 40.00 R57 radial. In fact, the 44.00 R57 tire was never produced. Instead, the big off-highway tire manufacturers went right up to an even larger 48/95 R57 radial, which was originally developed by Bridgestone. When equipped with the original 40-series tires, the first MT-4400 trucks could only manage a maximum 240-ton payload. In September 1995, the Caballo Mine (first operators of the model) fitted larger 48/95 R57 tires on the front wheels of two of their trucks as a test (see color section), which increased capacity to a respectable 260 tons. Later, Unit Rig would offer the new 46/90 R57 radial for trucks specified for 260-ton loads, and the 40.00 R57 size for 240-ton applications. Pictured in October 1995 is one of the Caballo Mine's MT-4400 Lectra Hauls. *Image by the author*

Terex Unit Rig placed its first MT-4400AC electric drive mining truck into service in July 2003. The MT-4400AC was much like the standard model of this very popular 260-ton hauler, but utilized a General Electric AC-drive system. In 2002 Unit Rig originally marketed the new truck as the TMT 260AC, to correspond with the model name changes that were being made across both the truck and hydraulic excavator product lines at Terex. But by the time the truck was ready for delivery, Terex Mining marketing decided to go back to the previous "MT" product line reference. Shown at Powder River Coal Company's Caballo Mine in July 2003 is the pilot MT-4400AC hauler. It is interesting to note that even though the model name of the new truck at this time was now MT-4400AC, its serial number tag mounted on the truck still read "TMT260." *Image by the author*

The payload capacity of the MT-4400AC was originally listed at 260 tons. This was made possible by big, beefy low-profile 50/80 R57 size radials on 57-inch rims. Optional tires offered were the 46/90 R57, and the 40.00 R57 (240-ton rating) series type. In late 2006 the truck's payload rating was revised once again to a nominal 240 tons for all tire options. Pictured on July 23, 2003 is the prototype MT-4400AC being loaded for the very first time at the Caballo Mine in the PRB of Wyoming. *Image by the author*

The MT-4400AC was offered with two 16-cylinder diesel engine choices; the MTU/DDC 16V4000 and the Cummins QSK60, both rated at 2,700 gross/2,487 net horsepower. The AC-drive system utilized a GE GTA-41 alternator, and GE B25 traction wheel motors. Maximum speed of the truck on level ground was 40 mph. This image of a MT-4400AC was taken at the Trans Alta Centralia Mine in Washington, in September 2006. *Urs Peyer*

The MT-4400 series has proven itself as one of the finest truck models Unit Rig ever engineered, with the MT-4400AC model being the best of the best in the product line. The truck's empty vehicle weight came in at approximately 377,000 pounds, with a fully loaded maximum gross weight of 865,000 pounds. Overall length was listed at 46 feet, 2 inches, with a maximum width of 26 feet, 2 inches. The MT-4400AC truck still lives on in the mining marketplace today as a product of Caterpillar, Inc., referred to as the Unit Rig MT4400. Thru December 2011 approximately 749 units had been sold of all model types of the MT-4400 (with 293 of that total being of the AC variety). *Terex*

Early on in the development of Unit Rig's ultra-hauler truck program the company was having difficulty with obtaining the technology for its drive system. Other manufacturers in the industry had already made long-term agreements with key suppliers (such as Komatsu with GE, and Liebherr with Siemens) of AC-drive engineering expertise necessary to bring a 320-plus-ton payload capacity hauler to life, making life for Unit Rig very difficult. Finally, in September 1997 Unit Rig announced that it had made an agreement with Power Conversion Systems (PCS) Division, an affiliate company of General Atomics of San Diego, California, for the development of an AC-drive system for their new ultra-hauler truck program. Early on in the development stage, the Unit Rig truck model was referred to as the MT-4800. But by mid-1998 the company had settled on the MT-5500 nomenclature. Shown on April 2, 2000 is the just completed pilot MT-5500 parked behind the main Terex Unit Rig plant in Tulsa, Oklahoma. *Image by the author*

Terex Unit Rig originally shipped its first MT-5500 to the Jacob's Ranch Mine in the Powder River Basin of Wyoming, for its initial "shake-down" field-test. The pilot truck would be up and running at the mine by August 2000. After a short stay there, it was shipped up to its permanent home at the Belle Ayr Mine just south of Gillette. The next five MT-5500 trucks built would all ship to the Grupo Mexico (GMexico) Cananea copper mine, located near Sonora, Mexico. Unfortunately, these trucks did not fit in well with the mine's operations, and all were eventually transferred to the Belle Ayr Mine to join the prototype unit. Shown at the Belle Ayr Mine in May 2002 is one of the original MT-5500 haulers equipped with an early generation "DT HI-LOAD" lightweight dump box. Problems with these bodies in the field forced Belle Ayr to eventually replace all of them with a more traditional design. *Image by the author*

During early development of the MT-5500 hauler program, the design had an original payload target of 340 tons (September 1998). By March 2000 the capacity had been increased to 360 tons. Power for the MT-5500 was supplied by either a MTU/DDC 16V4000, or a Cummins QSK60. Both diesel engines were rated at 2,700 gross/2,478 net horsepower each. The AC-drive system was supplied by General Atomics/PCS, including the alternator, W55 double-reduction traction motors, and retarding forced air grid system mounted on the upper deck. Standard tire size specified for the truck was 55/80 R63 radial tires mounted on 63-inch rims. This image of an MT-5500 was taken at the Belle Ayr Mine in May 2002. *Image by the author*

The 360-ton capacity MT-5500 was designed around a frame utilizing a 21-foot, 10-inch wheelbase. Its overall length was 48 feet, 6 inches (standard rock body), with a width of 31 feet. Empty vehicle weight of the MT-5500 was listed at approximately 478,000 pounds, with a maximum gross weight of 1,198,000 pounds. The MT-5500 had a slow start in the marketplace due mainly to its troublesome General Atomics/PCS AC-drive system. It would take several years to work out all of the design bugs out of the software, but eventually the trucks in the field all started to see great improvements in their overall availability. In 2004 Terex/Unit Rig offered an upgraded "B" version of the model with additional high-horsepower engine options. After Bucyrus completed its purchase of Terex Mining in February 2010, the MT-5500 became the Bucyrus MT5500AC Mining Truck. It would be reclassified as the Unit Rig MT5500 after Caterpillar acquired Bucyrus in July 2011. Total production of the MT-5500 by 2011 was only 49 units. Picture taken at the Belle Ayr Mine in September 2007. *Keith Haddock*

The largest mining truck ever designed by Terex/Unit Rig was the huge 400-ton capacity model MT-6300AC. The development process (like that of the MT-5500) was not a straight line to production. The project for this model actually started out in mid-2005 as the MT-5900AC. This design called for a truck with a payload capacity range of 360 to 390 tons, and was heavily based on the proposed MT-5500B program. But as the design of the new model progressed, its target payload capacity was raised to 400 tons, made possible by the use of 59/80 R63 radial tires mounted on 63-inch rims (with optional 56/80 R63 series type available). The MT-6300AC was officially announced by Terex Mining at the April 2007 BAUMA trade fair in Munich, Germany (first specifications dated March 2007), with the first actual prototype truck shipped to the oil sands mining area of Fort McMurray, Alberta, Canada, in February 2008. It would be delivered to Syncrude's Aurora Mine the following month for the beginning of its initial field-follow testing program. Shown is the pilot MT-6300AC on site in May 2008. *Keith Haddock*

The MT-6300AC is equipped with a unique patented hybrid, high-efficiency, long-tailed dump body featuring a curved floor, front, and canopy, with special runners that transfer weight to the frame. Engine of choice for the monster 400-tonner is the 20-cylinder MTU/DDC 20V4000 diesel, rated at 3,750 gross/3,492 net horsepower. Its AC-drive system is provided by General Atomics and its affiliate company Power Inverters (formerly PCS), which includes powerful IGBT microprocessor controls and W63 triple-reduction wheel motor assemblies. Empty vehicle weight of the massive truck is approximately 530,000 pounds, with a fully loaded maximum gross weight of 1,330,000 pounds. Overall length (with optional diagonal front ladder) is 51 feet, 1 inch, with a width of 31 feet, 10 inches. The MT-6300AC utilizes a chassis design featuring a box-beam front axle design, with a wheelbase of 21 feet, 10 inches. This pilot MT-6300AC equipped with a heated dump box was photographed in May 2008. *Keith Haddock*

The first operational fleet of Terex/Unit Rig MT-6300AC ultra-haulers was purchased by CITIC Pacific Mining (CPM) Management Pty Ltd, for use at the Sino Iron project (formerly referred to as the Cape Preston Iron Ore Project), located at Cape Preston, southwest of Karratha, in Western Australia's Pilbara region, with the first of fourteen 400-ton capacity trucks arriving on site for assembly in early July 2008. These trucks featured special heavy-duty, long-tailed ore dump bodies, and revised front ladders and steps to meet Australian mining safety requirements. After the completion of the purchase of Terex Mining by Bucyrus International in February 2010, the massive hauler continued on as the Bucyrus MT6300AC Mining Truck. Bucyrus would ship its first two MT6300AC trucks painted in their new corporate color scheme to Fort McMurray, Alberta, Canada, in December 2010. After Caterpillar purchased Bucyrus in July 2011, the two trucks in Fort McMurray were dismantled in August 2011. By 2011 only 21 of the MT-6300s had been placed into service. Pictured is one of CP Mining's MT-6300AC trucks in May 2009. *Allistair Cooke*

During the testing of the M-64 Ore Hauler, Unit Rig engineers were trying to come up with a Lectra Haul truck design for the coal industry. One of the earliest on record was from 1960 called the Lectra Haul BD-160. This ambitious electric-drive concept study called for a unitized design with identical two-axle powered front and rear drivetrains and cabs. Each axle utilized four tires giving the design a total of sixteen. Power was supplied by two engines, one in each end. Stated capacity was 160 tons. Another concept was the Lectra Haul BD-140 from 1964. The BD-140 was a more conventional layout utilizing an M-85 tractor unit and a 140-ton capacity bottom-dump trailer. But a lack of large enough tires and higher horsepower engines made the production of such a unit economically unfeasible at the time. In 1968 Unit Rig put forth a design proposal for a 180-ton capacity coal hauler utilizing the M-100 chassis as a tractor unit. Referred to as the BD-180, it would eventually go on to become the company's first production bottom-dump coal hauler. Shown at the Tulsa plant in February 1972 is the pilot Lectra Haul BD-180 (S.N. 51).

Principle design work on the Lectra Haul BD-180 took place throughout most of 1971, with the first prototype finished in February 1972. The first BD-180 was powered by a Detroit Diesel 12V-149TI engine, rated at 1,200 gross/1,070 net horsepower. Its electric-drive system utilized a Reliance UR22771 generator, and two Unit Rig-designed W-100 electric wheel assemblies featuring a modular arrangement, with a bolt-on gear package and two self-contained, bolt-on Reliance UR22770 electric traction motors. The prototype coal hauler was shipped to Consolidation Coal Company's (CONSOL) Glenharold Mine in Stanton, North Dakota, and would get its first load of coal on March 13, 1972. The next four BD-180s built by Unit Rig would feature Caterpillar D349 powered drivetrains utilizing a General Electric supplied GTA-18 generator and two 772 wheel motors. Horsepower ratings were the same as the Detroit Diesel unit. These would ship in September 1973 to AMAX Coal's Ayrshire Mine in Chandler, Indiana. Shown is the first BD-180 at the Glenharold Mine in March 1972.

In 1977 Unit Rig introduced a few improvements to the BD-180 Lectra Haul bottom-dump. The tractor on the unit now featured the new operator's cab that was being incorporated into all of the company's truck models, and a modified front end featuring a new ladder design and relocated air cleaners. This BD-180 (and all prier units) road on 30.00-51, 46PR tires mounted on 51-inch rims, with the suspension system utilizing the company's DYNAFLOAT rubber cushioned column type design. Payload was 180 tons, with a maximum 246-cubic-yard heaped capacity. Overall length was 78 feet, 2.75 inches, with an empty weight of approximately 232,300 pounds. The first six revised BD-180 coal haulers shipped from the factory in April 1977 to Carter Oil Company's Rawhide Mine (now owned by Powder River Coal Company), located just north of Gillette, Wyoming. Total number of BD-180 Lectra Hauls delivered into service was approximately 29 units. Pictured is one of the first of the new BD-180s assembled at the Rawhide Mine in July 1977.

The largest bottom-dump coal haulers produced by Unit Rig were its Lectra Haul BD-240/270 model series. The BD-240 utilized a tractor unit based on the Mk 36 chassis, pulling a large bottom-dump trailer capable of hauling a 354-cubic-yard coal load, with a nominal 240-ton payload capacity rating. The first unit built by Unit Rig shipped to Carter Mining Company's Caballo Mine (now operated by the Powder River Coal Co., a subsidiary of Peabody Energy), located just south of Gillette, Wyoming, in the Powder River Basin (PRB), and was assembled on site in December 1985 (pictured).

The BD-240 was certainly huge for its day, measuring 88 feet, 9 inches in length, with an overall width of 23 feet, 3 inches. Empty weight of the complete unit was approximately 390,950 pounds. Standard tire size utilized on the coal hauler was 36.00-51, 50PR series type for both the tractor and trailer. Its electric drive system utilized a GE GTA-22 alternator and two GE 776 wheel motors for the tractor unit. Pictured in December 1985 is the BD-240 in operation at the Caballo Mine.

During the early testing phase of the BD-240 at Caballo, it soon became evident to Unit Rig and mine engineers alike that the big Lectra Haul bottom-dump had far more payload potential. Unit Rig raised the side-boards on the trailer by a few inches and increased the nominal capacity payload rating to 270 tons. All of the units shipped after the first BD-240 were classified as BD-270 models. The BD-240/270 bottom-dump's Mk 36 tractor unit was offered with either a Detroit Diesel 16V-149TIB (the preferred choice), or a Cummins KTTA-50-C1600, both rated at 1,600 gross/1,450 net horsepower. To meet customer requirements, Unit Rig could ship a new tractor unit with the coal hauler, or rebuild one of the mine owner's existing Mk 36 rear-dumps and convert it into a tractor unit. Shown in October 1995 is the second bottom-dump unit to become operational at the Caballo Mine (1986), and the first to be referred to as a BD-270. *Image by the author*

The tractor specified for use on the Unit Rig BD-270 would change in 1990 with the introduction of the rear-dump MT-3600 Lectra Haul. When that rear-dump model became the MT-3600B in 1996, the tractor choice would also become a "B" version for the BD-270. Three engine packages were offered for the BD-270 equipped with the MT3600B tractor: a MTU/DDC 12V4000 (maximum optional power rating of 2,025 gross/1,885 net horsepower), and Cummins KTA/KTTA-50-C and QSK45 diesel engines (maximum optional power rating of 2,000 gross/1,860 net horsepower for each). The electric drive system utilized a GE GTA-22 alternator and two GE 788 wheel motors. Shown in October 1998 is a BD-270 Lectra Haul, equipped with an MT-3600B tractor unit, being loaded at Powder River Coal Company's Caballo Mine in the PRB. *Image by the author*

Though the Unit Rig BD-270 coal haulers sold in low numbers for the company, they were very reliable and productive in service for the mining operations that ran them. The last versions of the BD-270 haulers produced had an empty vehicle weight of approximately 414,627 pounds. Overall length of the unit equipped with the MT-3600B tractor unit was 90 feet, with a width of 24 feet, 3 inches. In 2003 Unit Rig offered a proposed 380-ton capacity BD-380AC coal hauler, but none were ever built. Pictured in June 2000 is a BD-270 operating at Peabody Energy's Lee Ranch Mine, located near Grants, New Mexico. Note the absence of the "Lectra Haul" trade name, which had been quietly phased out of use by late 1999. *Image by the author*

In late 1974 Unit Rig released preliminary specifications for a two-axle, unitized bottom-dump coal hauler referred to as the BD-145 Lectra Haul. The BD-145 utilized a rear-mounted, lift-out engine module for better weight distribution and greater traction and control on adverse grades. Its front ackerman-type, turret steering design allowed the coal hauler to make full 90-degree turns, and featured the DYNAFLOAT rubber cushion column type suspension system. The prototype BD-145 was equipped with a Detroit Diesel 12V-149TI engine, rated at 1,200 gross/1,070 net horsepower, and utilized a Unit Rig-designed electric drive system consisting of a Reliance UR RTG-101 traction generator and two W-100 modular wheel motor assemblies. Standard tires were 30.00-51, 40PR type. Empty weight was 175,768 pounds, with a maximum payload capacity of 145 tons in its 187-cubic-yard heaped hopper. Length of the prototype unit was 53 feet, 6 inches. The BD-145 was shipped to Arch Minerals' Southwestern Illinois Coal Corporation's Captain Mine, located near Percy, Illinois, in July 1977 for field-testing (pictured at the mine in the fall of 1977).

After less than a year of testing at the Captain Mine, the prototype BD-145 was shipped back to Unit Rig's Tulsa plant for a complete engineering evaluation. The result of the redesign program was the BD-30 Lectra Haul. The BD-30 featured a semi-monocoque frame with a redesigned hopper construction, which looked much different than the previous BD-145 prototype. The redesigned coal hauler was officially unveiled at the AMC show in Las Vegas, in October 1978. The show BD-30 featured a Cummins KTA-2300 equipped engine module (1,200 gross/1,075 net horsepower), and was equipped with General Electric's SEPEX drive-system, consisting of a GE 603K3 generator and two GE 772VS wheel motors mounted in the rear. Payload capacity of the BD-30 was 160 tons (194 cubic-yards heaped), with an empty vehicle weight of approximately 218,000 pounds. The show unit was also fitted with Firestone 33-59.5, 46PR tires mounted on special Firestone developed one-piece drop center rims. Picture taken at the AMC show in October 1978. *Author's Collection*

In February 1979 Unit Rig shipped three production BD-30 HD Lectra Haul two-axle bottom-dump coal haulers to Kerr-McGee's Jacobs Ranch Mine in the PRB coal mining region of Wyoming. The BD-30 HD utilized more robust GE electric drive components and controls, with higher ply rated tires. Standard tires specified were now 30.00-51, 46PR series type, mounted on 51-inch rims. Empty vehicle weight of this version of this unique coal hauler tipped the scales at 234,000 pounds, but total payload capacity remained at 160 tons. Overall length of the BD-30 HD was listed at 56 feet, with an overall width of 20 feet, 10 inches. Unfortunately, the severe worldwide economic recession of the early 1980s ended all interest in the marketplace for the company's unitized coal hauler concept, causing Unit Rig to abandon the program. Shown on site in March 1979 is one of the BD-30 HD Lectra Hauls.

The Haulpak Model 120 series of off-highway trucks were stellar performers for WABCO and sold quite well in large mining operations worldwide. This 120A is shown working at Kennecott Copper's Ruth Mine in Ely, Nevada, in September 1966.

Shown on display at Komatsu's Peoria assembly plant in December 1996 is this beautifully restored LW-32. This LW-32 was actually the fifth Peoria-built off-highway truck produced, and was delivered to Utah Construction & Mining Company's iron ore mine near Cedar City, Utah, in April 1958. It was officially retired by the mine in 1983 with 80,153 hours of operation. The Haulpak distributor in that territory, Rocky Mountain Machinery Co., would have the truck restored for Dresser to be put on display at the AMC show held in Las Vegas, Nevada, in October 1986. After the show the truck was placed into storage and would eventually be donated by the distributor to Komatsu and shipped back to its place of birth in Peoria to be completely restored again, which was completed in 1996. As of 2011, it is currently in storage at the back of the Peoria assembly plant.

Shown is a 150-ton capacity WABCO Model 150Bw Haulpak being commissioned at the Twin Buttes Mine near Tucson, Arizona, in December 1972.

The Model 170D Haulpak was originally introduced by WABCO in 1980. It was rated as a 170-ton capacity hauler. Pictured in August 1985 is a 170D in new Dresser-WABCO colors.

When the WABCO Model 3200 Haulpak was originally introduced in late 1971, it was only rated as a 200-ton capacity hauler. By 1973 this had been increased to 235 tons. Image taken in October 1971.

Shown in May 1973 at the Peoria plant is one of the first 3200 Haulpak trucks to feature an early design of an exhaust heated, side-boarded dump box, which increased payload capacity to 235 tons.

In the summer of 1974, WABCO officially announced the introduction of the Model 3200B Haulpak off-highway truck featuring an increased payload capacity of a nominal 235 tons. Pictured in October 1975 is a "B" unit working at the Lornex Mining Company's Logan Lake Mine in British Columbia, Canada.

By 1978 WABCO had upgraded the capacity of the 3200B to 235-260 tons capacity, depending on the use of dump body bed liners. The 3200B image shown was taken in June 1976 at Duval Copper's Sierrita Mine located just south of Tucsan, Arizona.

The WABCO Model 200B Haulpak from 1969 was the direct result of the Model 160A rear-dump tractor/trailer program. The tractor unit of the 200B was based largely on the design of the 120B/150B chassis. This image of the first 200B Haulpak was taken at Duval Copper's Sierrita Mine in October 1969.

Pictured in October 1971 is the improved version of the WABCO Model 200B Haulpak. Payload capacity of this unusual 3-axle, four-wheel diesel-electric drive hauler was 200 tons.

WABCO produced a good number of bottom-dump coal haulers and were some of the very best in the industry at the time. Shown in May 1967 is a 120-ton capacity 75B/120 Haulpak Coal Hauler painted in the striking colors of Peabody Coal Company.

The only mine site to take delivery of the unique looking unitized WABCO Model 170 Coalpak trucks when new was the Jacobs Ranch Mine, located in the Powder River Basin of Wyoming. Image taken in October 1979.

The mechanical-drive Komatsu 330M Haulpak was rated at a 100-ton capacity when it was originally introduced in 1991. In early 2000 its designation was changed to the HD785-5 and the Haulpak name was dropped. Photo taken in June 1997. *Image by the author*

Shown with a full 240-ton coal load is a Komatsu 330M Haulpak tractor pulling MEGA Magnum Tandem CH120 coal bottom-dump trailers, operating at North American Coal's San Miguel Lignite Mine near Jourdanton, Texas, in December 1997.

The Komatsu HD1500-7 (along with its predecessor, the HD1500-5) can trace its origin back in the North American marketplace to the Komatsu 530M Haulpak off-highway truck originally released in 1996. Photo taken in April 2008. *Komatsu America Corp.*

The Komatsu 730E was originally introduced in 1995 under the Haulpak name, and was rated with a maximum payload capacity of 212 tons (later lowered to 205 tons). Pictured in April 1998 is a 730E Trolley truck equipped with a front-mounted electric pantograph system which was in use at the Barrick Goldstrike Mine in Nevada at the time. Payload capacity of the trolley-equipped 730E was limited to 202 tons.

Dresser officially took the wraps off of its new 240-ton capacity, diesel-electric drive 830E Haulpak off-highway truck at the 1988 MINExpo show held in Chicago, Illinois. Image taken on April 26, 1988. *Author's collection*

The first two customers of the new Dresser 830E Haulpak were both coal mining operations operating in the Powder River Basin of Wyoming—Kerr-McGee Coal Corp. and Cordero Mining Co. Pictured in November 1988 is one of the first Cordero units fitted with a huge volume capacity coal body.

The Dresser 830E was perfectly matched to the largest loading shovels of the day, such as this electric Demag H485S front shovel working at the Asarco Ray Mine in September 1996. *Image by the author*

Once Komatsu became sole owner of Dresser's mining hauler product lines, the Dresser name would be removed from all of the trucks, such as the 830E. By the end of 1999 the Haulpak reference would also be quietly discontinued. But today's Komatsu 830E still has the heart of a "Haulpak" and is considered one of the finest diesel-electric drive 240-255-ton class capacity haulers ever built. Photo taken in July 2003. *Image by the author*

Komatsu first introduced their 830E-AC electric drive truck at the September 2004 MINExpo. Since then it has proven itself as a highly productive hauler for mining operations that can benefit from the advantages AC-drive technology offers. *Komatsu America Corp.*

At the 2008 MINExpo held in Las Vegas, Komatsu officially introduced a new model line to the press in the form of the 860E-1KT. This truck featured the company's new Komatsu Drive System AC technology and a factory-installed trolley-capable option. Prototype testing of the new truck design was conducted in 2007 and 2008 at Kumba Iron Ore's Sishen Mine. Photo taken on September 23, 2008. *Image by the author*

The Komatsu 860E-1KT (1K with no trolley option) is powered by a Komatsu SSDA16V160 Tier 2 rated at 2,700 gross/2,550 net horsepower. Payload capacity is a healthy 280 tons. The first fleet of 860E-1KT haulers began full operations in 2008 at Kumba Iron Ore's Sishen Mine in Northern Cape, South Africa. Photo taken in June 2008. *Komatsu America Corp.*

Komatsu Dresser officially announced its new AC electric drive 930E Haulpak mining truck in May 1995. The company's first field unit would haul its first payloads at Fording River Coal, located near Elkford, British Columbia, Canada. Pictured is the 310-ton capacity 930E at Fording in October 1997. *Image by the author*

As the 930E Haulpak program commenced, its payload capacity was quickly raised to 320 tons. Pictured in September 1996 is the third 930E truck built operating at the Asarco Ray Mine, located near Hayden, Arizona. *Image by the author*

In 1999 the Haulpak trade name was removed from Komatsu's quarry and mining truck product lines, including the 930E. In late 1999 a new version of the company's ultra-hauler, the 930E-2, was introduced equipped with new larger, low-profile 53/80R63 tires. This new model was initially referred to as the "Bigfoot." Photo taken in November 2004. *Image by the author*

Komatsu released an updated 930E-4 version of its very successful 320-ton capacity mining truck in early 2007, featuring many improvements in regard to efficiency and reliability issues. Power and payload ratings were un-changed from the previous model. *Komatsu America Corp.*

The Komatsu 930E series of trucks are some of the most productive 300-plus-ton capacity, diesel-electric drive mining haulers ever built, and are in service at large mining operations the world over. In January 2011, the 1,000th production 930E (all model types) shipped from Komatsu's Peoria, Illinois, manufacturing plant. Shown operating in the oil sands in northern Alberta in July 2009 is a new 930E-4 equipped with a special curved canopy dump body custom fabricated by the local Komatsu dealer. *Keith Haddock*

Komatsu's first pilot program of its FrontRunner "Autonomous Haulage System" (AHS) was conducted at Codelco Norte's Radomiro Tomic copper mine in Northern Chile. Five trucks were tested in an isolated area of the mine beginning in December 2005 through 2007. The success of this pilot program led to the first operational fleet of eleven 930E-AT haulers being commissioned at Codelco's Gaby Mine (since renamed the Gabriela Mistral Mine) in early 2008. The next fleet of 930E-AT trucks equipped with AHS was shipped to Rio Tinto's West Angelas Mine, East Pilbara operation in Western Australia. The first autonomous amount of overburden moved by a truck at the West Angelas Mine was made on December 18, 2008. Full-time day and night production of the AHS fleet started in April 2009. In November 2011 Komatsu Ltd. and Rio Tinto signed a Memorandum of Understanding for delivery of at least 150 AHS trucks into Rio Tinto Iron Ore Pilbara operations in Western Australia by the end of 2015, with initial deployment commencing in 2012. This striking shot of a 930E-AT was taken at the West Angelas Mine in May 2009. *Image Courtesy Rio Tinto*

During the 2000 MINExpo show held in Las Vegas (October 9-12), the worldwide 930E hauler fleet hit the magic milestone of 2 million operating hours. Komatsu celebrated the special occasion by tossing balloons from the upper deck of the new 930E-2SE mining truck to on-lookers below. *Image by the author*

With the introduction of the 930E-4 in 2007, the far more powerful 930E-3SE was also upgraded to a 4SE designation. Power for the 930E-4SE is provided by an 18-cylinder Komatsu SSDA18V170 diesel engine rated at 3,500 gross/3,429 net horsepower. Shown in June 2008 is a 930E-4SE equipped with a DT Hi-Load dump body built by Dicsa Tricon of Santiago, Chile. *Komatsu America Corp.*

On March 1, 2010, Komatsu commemorated the 930th 930E built (all version types) at its Peoria manufacturing facilities. Shown are some of the proud members of Local Lodge No. 158 of the International Brotherhood of Boilermakers, Iron Ship Builders, Blacksmiths, Forgers and Helpers, that have built these massive trucks over the years. The 930th truck (930E-4SE) was shipped to the Collahuasi Copper Mine in northern Chile. *Komatsu America Corp.*

Shown in July 2006 is the pilot Komatsu 960E-1K in Fort McMurray, Alberta, Canada, just before its delivery to Suncor to start its long-term field trial program. The New Komatsu Drive System installed in this truck was developed in cooperation between Komatsu and Siemens. *Komatsu America Corp.*

The Komatsu 960E-1K is powered by a Komatsu SSDA18V170 diesel engine, the same powerplant found in the 930E-4SE and standard 960E-1. Power output is rated at 3,500 gross/3,346 net horsepower. Payload capacity is rated at 360 tons. The 960E-1K was officially released for worldwide sales in 2010. Image date is July 2006. *Komatsu America Corp.*

Komatsu first started its field-follow testing program on its new 360-ton capacity 960E-1 ultra-hauler in June 2005. In early 2008 the truck was officially released for sale. Pictured is the 960E-1 that was on display at the September 2008 MINExpo in Las Vegas. *Image by the author*

The prototype Unit Rig M-64 ore hauler from 1960 was the company's first mining truck to carry the "Lectra Haul" product name. Though the articulated frame design would not see future production, its General Electric drive system would—and in a big way, changing the face of diesel-electric drive mining trucks in the industry forever. Pictured is the M-64 in late January 1960 being previewed at the Standard Industries, Inc. quarry, located east of Tulsa, Oklahoma.

The success of the GE drive system in the M-64 prototype, utilizing electric traction motors in the wheel assemblies, would form the drivetrain basis for Unit Rig's first true production mining truck, the Lectra Haul M-85. Pictured is the first M-85 in August 1963 at Kennecott Copper's Chino Mine in New Mexico.

Unit Rig kept a close eye on its new Lectra Haul M-85 ore hauler during its early months in the field, and saw areas where the design could be improved. These changes were incorporated immediately in an updated version in 1964. Pictured at the Tulsa plant in May 1964 is the first of these improved Lectra Haul M-85 trucks.

Unit Rig's first two Lectra Haul M-100 models built both featured turbine-electric drivetrains. The first was equipped with an IH Solar turbine engine, and the second utilized a GE unit. Shown is the Lectra Haul M-100 equipped with the GE turbine engine being christened at the Unit Rig display at the AMC show in Las Vegas, Nevada, in October 1965.

The turbine equipped Lectra Haul M-100 prototypes would not go into full production, but that would not diminish the popularity of the standard diesel-electric drive version of the model, which would go on to be a huge success for the company in the marketplace with approximately 1,092 units delivered into service by 1981. Pictured in 1967 is one of the upgraded diesel-electric M-100 haulers featuring a redesigned front ladder arrangement.

This fine looking Lectra Haul M-100 was taken at the Unit Rig Tulsa plant in late 1967, just before delivery to Rio Tinto's Borax Mine in Boron, California. Borax took delivery of their first M-100 in mid-1966, and would eventually operate a fleet of twelve of the 100-ton capacity trucks.

Unit Rig's first three models of Lectra Haul trucks all utilized a chassis based on a 15-foot wheelbase. The largest of these was the M-120 introduced in early 1968. By 1974 a version based on a 17-foot wheelbase chassis design was released, referred to as the M-120.17. All early versions of the model were reclassified as M-120.15 models to denote the use of a 15-foot wheelbase. Shown is a Lectra Haul M-120.17 working at the Petrotomics uranium mine (Division of Getty Oil), located in Shirley Basin, Wyoming, circa 1977.

Unit Rig really took the mining industry by surprise in 1968 with the unveiling of their 200-ton capacity Lectra Haul M-200. The M-200 was the first two-axle, diesel-electric drive hauler to achieve this payload capacity, which was considered huge for its day. Pictured in the summer of 1968 is the pilot M-200 at the Tulsa plant, just before delivery to Las Vegas for its world debut at the AMC show.

The first operational fleet of Lectra Haul M-200 haul trucks was commissioned in early 1969 at Kaiser Coal Ltd.'s (Kaiser Resources) mining operation, located near Sparwood, British Columbia, Canada. Shown in April 1969 are some of the first M-200 trucks assembled on site. Total number of M-200 units operated by Kaiser was 22.

Problems with severe tire wear on the Lectra Haul M-200 would finally be solved with the introduction of the larger 40.00-57, 60PR-sized rubber, replacing the previous 36.00-51 series type. This new tire package was featured on the display M-200 at the 1971 AMC show in Las Vegas. Pictured in 1972 is an M-200 fitted with the new larger tire and wheel package.

Shown in June 1972 is the pilot 170-ton capacity Mk 36 Lectra Haul off-highway mining truck during final assembly at Unit Rig's Tulsa plant. This truck would ship from the factory in August 1972 to the Cyprus Pima Mine, located south of Tucson, Arizona, for the start of its field-follow testing program.

Lectra Haul trucks wearing a Canadian maple leaf decal on their radiator sides (such as this picturesque Mk 36 from the early 1980s) were assembled at one of Unit Rig & Equipment Co. (Canada) Ltd. facilities. Founded in 1965, the Canadian subsidiary company purchased its Niagara Falls facilities in 1970, followed by the Welland plant three years later. In 1977 it opened the Stevensville, Ontario, plant to take over assembly and warehousing from the Niagara operation, which in turn would take over fabrication operations previously handled by the now outdated Wellman facility.

The Unit Rig Mk 36 Lectra Haul is considered one of the greatest large mining truck designs of the 1970s. Though early trucks in the field were plagued by frame problems, this did not seem to diminish the truck's overall production benefits to the mining companies that operated them. Shown in May 1980 is a Mk 36 operating at Erie Mining Company's iron mine in Minnesota.

The 205-ton capacity Unit Rig MT-3700AC mining truck was a solid performer for Terex, and would continue on as a Bucyrus model in 2010. It was also the first truck model delivered to a customer in the new Bucyrus colors (May 2010). After Caterpillar's acquisition of Bucyrus in July 2011, it would once again become a "Unit Rig" model. Image date is March 2006. *Terex*

The first MT-4400 Lectra Haul trucks delivered in 1995 had various tire problem issues that were temporarily addressed with the installation of larger 48/95 R57 tires on the front wheels. Shown in October 1995 is one of these modified MT-4400 units working at the Caballo Mine, located in the Powder River Basin coal mining region of Wyoming. *Image by the author*

The MT-4400 series of mining trucks are considered some of the best all-around designs Unit Rig ever built. The last version released by the company, the MT-4400AC, was first placed into service in July 2003. Pictured is an MT-4400AC in January 2006. *Terex*

Shown in April 2000 is the prototype Unit Rig MT-5500 at the Tulsa plant during an employee open-house event, just before delivery to its first field-tests in the Powder River Basin of Wyoming. *Image by the author*

Terex Mining first announced its massive MT-6300AC hauler in 2007. Originally referred to as the MT-5900AC during development (2005), it became the MT-6300AC with the availability of larger 59/80 R63 tires. Shown at the September 2008 MINExpo in Las Vegas is the mighty 400-ton capacity MT-6300AC, soon to be delivered to CP Mining in Western Australia, after the show. *Image by the author*

The first mining company to purchase Terex Mining's Unit Rig MT-6300AC ultra-haulers was CITIC Pacific Mining Management Pty Ltd, for use at its Sino Iron project, located in the Pilbara region of Western Australia. The prototype truck tested in the oil sands remained the property of Terex at the time of its initial delivery in 2008. Shown is one of CP Mining's MT-6300AC trucks in May 2009. *Allistair Cooke*

Though the Lectra Haul BD-180 was in production from 1972 to 1984, only approximately 29 units were ever delivered into service. Shown is the prototype BD-180 at CONSOL's Glenharold Mine in March 1972.

Unit Rig's big Lectra Haul BD-240/270 bottom-dump coal haulers were the largest ever designed and built by the company. The first of these giants delivered to the Caballo Mine, in the Powder River Basin of Wyoming in December 1985, was the only BD-240 built. After this, all examples of this coal hauler were considered BD-270 models. Pictured is a BD-270 at the Caballo Mine in October 1995.

Unit Rig's first unitized coal bottom-dump design to be placed into field trials was the Lectra Haul BD-145. The prototype unit shipped from the Tulsa plant in July 1977, and it is shown here shortly after delivery to Arch Minerals' Captain Mine, located near Percy, Illinois. This prototype would eventually lead to the production of the BD-30 model line.

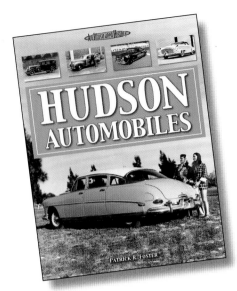  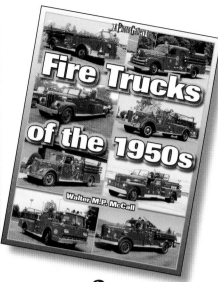

More Great Titles from Iconografix

All Iconografix books are available from specialty book dealers and bookstores worldwide, or can be ordered direct from the publisher. To request a FREE CATALOG or to place an order contact:

Iconografix
Dept BK
1830A Hanley Road
Hudson, WI 54016

Call:
(800) 289-3504
715-381-9755

Email: info@iconobooks.com

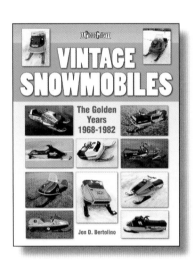 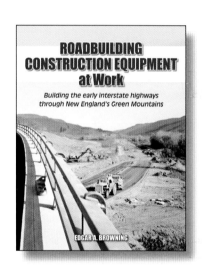

To see our complete selection of over 7,000 titles, visit us at:
www.enthusiastbooks.com
We would love to hear from you!

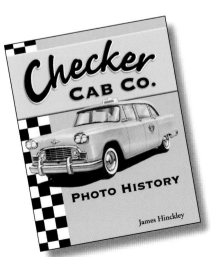 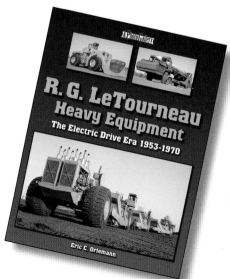